Math Enrichment: GRADE 5

Math Enrichment: GRADE 5

Students need to develop a solid sense of numbers and build the competence and confidence to compute, estimate, reason, and communicate to solve real-life problems. *Math Enrichment* engages students in the mathematics of real life–by reinforcing concepts, skills, and strategies that directly relate to daily living. Varied mathematical situations and intriguing content-area connections help students extend and enrich their understanding.

The activities in *Math Enrichment* complement and enhance your mathematics program. They spark both the teacher's and student's creativity and understanding. You will find this variety of activities engaging and challenging to *all* students!

ORGANIZATION

Math Enrichment is organized into eight units that focus on the essential areas of mathematics: number, problem solving, logic, patterns, algebra, measurement, geometry, and probability and statistics.

NUMBER
Includes puzzles, mazes, games, tables, and activities that reinforce addition, subtraction, multiplication, division, decimals, fractions, and percent.

PROBLEM SOLVING
Includes graphs, stories, maps, tables, and activities that reinforce the basic operations and estimation.

LOGIC
Includes stories, games, charts, puzzles, and activities that reinforce money, odd and even numbers, and problem solving.

PATTERNS
Includes charts, stories, puzzles, and visuals that reinforce skip-counting, multiplication, division, fractions, and geometry.

ALGEBRA
Includes puzzles, recipes, and real-life problem solving that reinforce decimals, the basic operations, proportion, and measurement.

MEASUREMENT
Includes using astronomy, graphic design, travel schedules, a water clock, maps, recipes, and graphs that reinforce distance scale, time, area, weight, temperature, currency, and weather.

GEOMETRY
Includes area, fractions, angles, networks, perimeter, and solid figures.

PROBABILITY AND STATISTICS
Includes members of, descriptions of, and identifying sets, Venn diagrams, and logic boxes.

Math Enrichment: GRADE 5

USE

The activities in this book are designed for independent use by students who have had instruction in the specific skills covered in the lessons. Copies of the activity sheets can be given to individuals or pairs of students for completion. When students are familiar with the content of the worksheets, they can be assigned as homework.

To begin, determine the implementation that fits your students' needs and your classroom structure. The following plan suggests a format for this implementation.

1. **Administer** the Assessment Tests to establish baseline information on each student. These tests may also be used as post-tests when students have completed a unit.

2. **Explain** the purpose of the worksheets to the class.

3. **Review** the mechanics of how you want students to work with the activities. Do you want them to work in pairs? Are the activities for homework?

4. **Introduce** students to the process and purpose of the activities. Work with students when they have difficulty. Give them only a few pages at a time to avoid pressure.

ADDITIONAL NOTES

1. Parent Communication. Send the Letter to Parents home with students.

2. Student Communication. Encourage students to share the Letter to Students with their parents.

3. Bulletin Board. Display completed worksheets to show student progress.

4. Student Progress Chart. Duplicate the grid sheets found on pages 6-7. Record student names in the left column. Note date of completion of each lesson for each student.

5. Curriculum Correlation. This chart helps you with cross-curriculum lesson planning.

6. Have fun! Working with these activities can be fun as well as meaningful for you and your students.

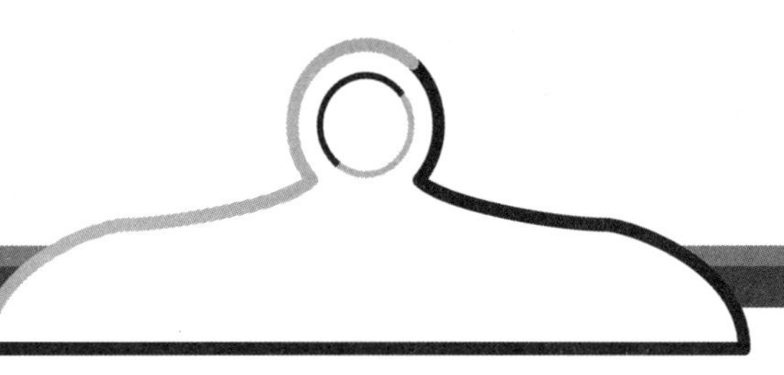

Dear Parent:

During this school year, our class will be working with mathematical skills. We will be completing activity sheets that provide enrichment in the areas of number, problem solving, logic, patterns, algebra, measurement, geometry, and probability and statistics.

From time to time, I may send home activity sheets. To best help your child, please consider the following suggestions:

- *Provide a quiet place to work.*
- *Go over the directions together.*
- *Encourage your child to do his or her best.*
- *Check the lesson when it is complete.*
- *Go over your child's work, and note improvements as well as problems.*

Help your child maintain a positive attitude about mathematics. Let your child know that each lesson provides an opportunity to have fun and to learn. If your child expresses anxiety about these strategies, help him or her understand what causes the stress. Then talk about ways to eliminate math anxiety.

Above all, enjoy this time you spend with your child. He or she will feel your support, and skills will improve with each activity completed.

Thank you for your help!

Cordially,

Dear Student:

This year you will be working in many areas in mathematics. The activities are designed for fun and for real-life applications. You will complete puzzles and mazes, break codes, solve mysteries, make tables and charts, and read maps. You will get to work with recipes, puzzles, graphs, planets and stars, stories, maps, and money. These activities will show you fun ways to practice mathematics!

As you complete the worksheets, remember the following:

- *Read the directions carefully.*
- *Read each question carefully.*
- *Check your answers after you complete the activity.*

You will learn many ways to solve math problems. Have fun as you develop these skills!

Sincerely,

STUDENT PROGRESS CHART

STUDENT NAME	UNIT 1													UNIT 2													UNIT 3										UNIT 4									
	NUMBER													PROBLEM SOLVING													LOGIC										PATTERNS									
	13	14	15	16	17	18	19	20	21	22	23	24	25	26	27	28	29	30	31	32	33	34	35	36	37	38	39	40	41	42	43	44	45	46	47	48	49	50	51	52	53	54	55	56	57	58

STUDENT PROGRESS CHART

STUDENT NAME	UNIT 5						UNIT 6														UNIT 7												UNIT 8		
	ALGEBRA						MEASUREMENT														GEOMETRY												STATISTICS		
	59	60	61	62	63	64	65	66	67	68	69	70	71	72	73	74	75	76	77	78	79	80	81	82	83	84	85	86	87	88	89	90	91	92	93

CURRICULUM CORRELATION

	Social Studies	Food and Nutrition	Music	Science	Physical Education	Art	Business
Unit 1: Number	17, 22	25		20, 24, 25		13	14, 23
Unit 2: Problem Solving	26, 30, 31, 33, 36, 38	32	28	35, 37		29, 34	26, 28, 38
Unit 3: Logic	39, 41, 44, 45	48		47, 48	43, 46	42	
Unit 4: Patterns	55, 56					56, 57, 58	
Unit 5: Algebra	59	63		64			60
Unit 6: Measurement	67, 68, 70, 71, 72, 75	74		65, 73, 76, 77		66, 67	70, 76, 78
Unit 7: Geometry	80, 82				84	79, 86, 90	
Unit 8: Probability & Statistics		91			92, 93, 94		91

Name ______________________ Date ______________________

Assessment: Units 1 and 2

1. Place the numbers in order, from least to greatest.

0.11 0.093 0.112 0.003 0.092

a. 0.11, 0.112, 0.003, 0.092, 0.093

b. 0.003, 0.092, 0.093, 0.11, 0.112

c. 0.112, 0.11, 0.093, 0.092, 0.003

d. 0.003, 0.112, 0.11, 0.092, 0.093

2. The pilot of a small jet is planning how to load the cargo into his jet. If the cargo is not balanced properly, the plane will not be able to fly. The pilot must arrange 1,284 pounds of cargo in the jet's three cargo compartments. Complete the pilot's work.

$\frac{1}{4} \times 1{,}284 = 321$ pounds in Compartment 1

$\frac{1}{4} \times 1{,}284 = 321$ pounds in Compartment 2

$\frac{1}{2} \times 1{,}284 =$ ______ pounds in Compartment 3

a. 321 **b.** 2,568 **c.** 642 **d.** not enough information given

3. The new music group, The Backfires, made a music video of their hit single. Their video was 2 minutes 23.3 seconds longer than the current Drones video. The Drones video lasts 5 minutes 30 seconds. What number sentence expresses the length of the Backfires' video?

a. 5 min 30 s + 2 min 23.3 s = *n*

b. 5 min 20 s - 2 min 23.3 s = *n*

c. 30 s + 23.3 s = n

d. 2 min 23.3 s × 5 min 30 s = *n*

4. Use the circle graph to solve.
Of the large predators, 8 are tigers. How many of the large predators are not tigers?

a. 692 **b.** 22 **c.** 202 **d.** 622

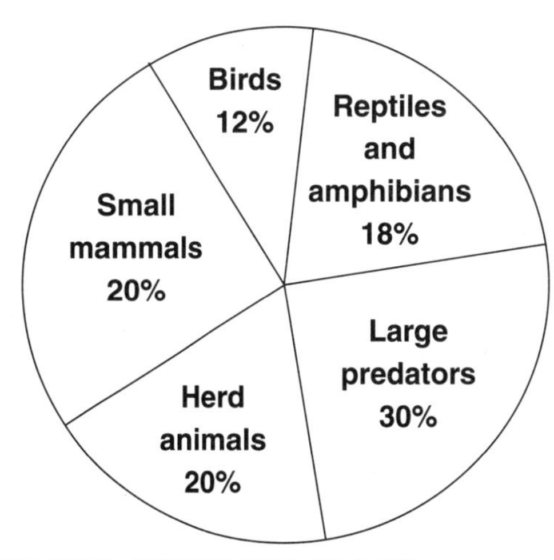

ANIMALS AT THE COLORADO CANYON INSTITUTE (700 animals in all)

Name ______________________ Date ______________________

ASSESSMENTS

Assessment: Units 3 and 4

1. LaToya has two dollars' worth of nickels, dimes, and quarters. She has a total of twenty coins. She has the same number of dimes as quarters. How many of each coin does she have?

a. 4 quarters, 4 dimes, 12 nickels

b. 7 quarters, 2 dimes, 1 nickel

c. 3 quarters, 3 dimes, 20 nickels

d. 10 quarters, 5 dimes, 5 nickels

2. Mortica is as old as her brothers Luke and Wesley together. Last year, Luke was twice as old as Wesley. Two years from now, Mortica will be twice as old as Wesley. How old are they?

a. Mortica: 8; Luke: 5; Wesley: 3

b. Mortica: 20; Luke: 10; Wesley: 10

c. Mortica: 6; Luke: 4; Wesley: 2

d. none of the above

3. What number is missing from the sequence?

$7\frac{1}{2}$, 6, $4\frac{1}{2}$, ____, $1\frac{1}{2}$,...

a. 2 **b.** $2\frac{1}{2}$ **c.** 3 **d.** $3\frac{1}{2}$

4. Use this progression to answer the question.

$\frac{1}{2}$, 1, $1\frac{1}{2}$, 2, $2\frac{1}{2}$, 3

What is the difference between each term in the progression?

a. $3\frac{1}{2}$

b. $10\frac{1}{2}$

c. 1

d. $\frac{1}{2}$

Name ______________________ Date ______________________

Assessment: Units 5 and 6

1. Find the number value of n in the multiplication problem.

$8 \times n = 6 \times 8$

a. 4 **b.** 6 **c.** 8 **d.** 48

2. What basic property of multiplication is demonstrated in problem 1?

a. Identity Property **b.** Distributive Property

c. Associative Property **d.** Commutative Property

3. Four students work in a bookstore on weekends to earn extra money. They made a table to show how many hours they spend doing different tasks.

Name	Total hours worked	Fraction of total hours		
		Working cash register	Unpacking books	Shelving books
Ellen	$5\frac{1}{4}$	$\frac{1}{6}$	$\frac{1}{6}$	$\frac{2}{3}$
Harold	$8\frac{1}{3}$	$\frac{1}{5}$	$\frac{2}{5}$	$\frac{2}{5}$
Lucia	9	$\frac{1}{3}$	$\frac{1}{6}$	$\frac{1}{2}$
Steve	$12\frac{1}{2}$	$\frac{3}{8}$	$\frac{1}{8}$	$\frac{1}{2}$

Who spends the least amount of time unpacking books?

a. Ellen **b.** Harold **c.** Lucia **d.** Steve

4. When the student councils at Pine Ridge School and Marble Creek School had a 4-week newspaper drive, they used a double-bar graph to show the number of pounds they collected each week.

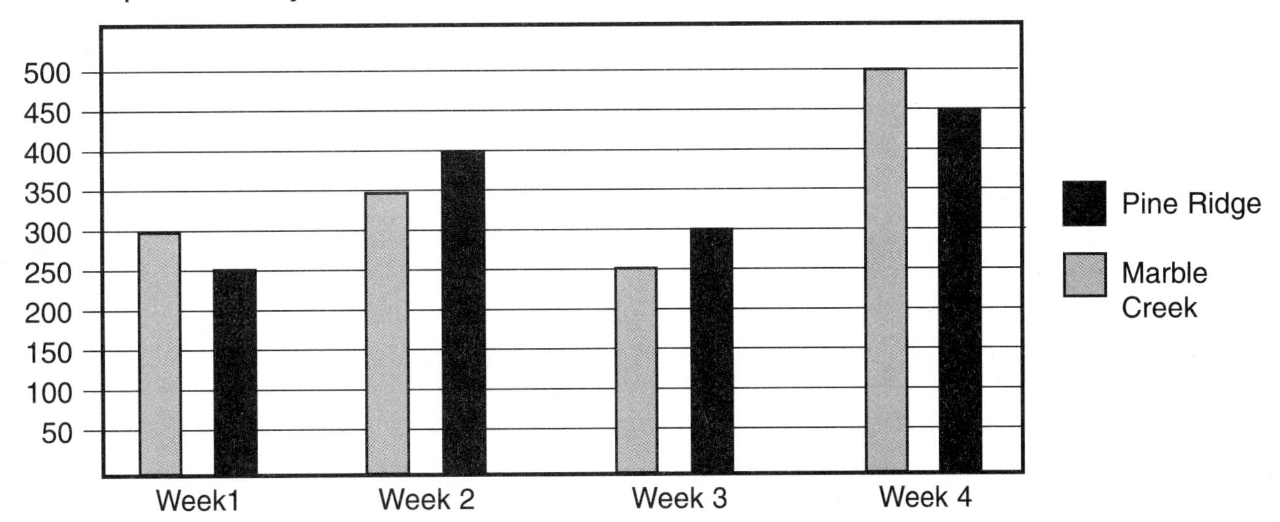

How many pounds of newspaper were collected by both groups in Week 3?

a. 550 lbs **b.** 250 lbs **c.** 300 lbs **d.** 650 lbs

Name ______________________ Date ______________________

Assessment: Units 7 and 8

1. Each wall in Ms. Wu's living room is 2.42 meters wide and 3.07 meters high. Estimate the area of the 4 walls in the living room.

a. 20 square meters

b. 24 square meters

c. 10 square meters

d. 8 square meters

2. What is the geometric description of a city on a road map?

a. point

b. line segment

c. parallel lines

d. plane

3. 

How many arcs connect B and C?

a. 0 **b.** 1 **c.** 2 **d.** 3

4. Use the table to answer the question.

TENNIS TABLE

Name	G (Girl) B (Boy)	Years playing	Age	Name	G (Girl) B (Boy)	Years playing	Age
Jacqueline	G	5	11	Enrico	B	3	12
Roscoe	B	4	10	Lynn	G	1	10
Shivaji	B	1	11	Ali	B	5	11
Sabrina	G	3	14	Juan	B	2	13

To the nearest year, what is the average age of all the tennis players?

a. 23 **b.** 12 **c.** 13 **d.** 92

Name ______________________ Date ______________________

NUMBER

Unit 1: Rounding and Estimating Sums

Cherise, John, Mabel, and Raoul entered Photo Fair's contest. They have to estimate which combination of a camera, tripod, camera bag, and two lenses comes closest to $1,000 without going over that figure. The person who comes closest wins the equipment.

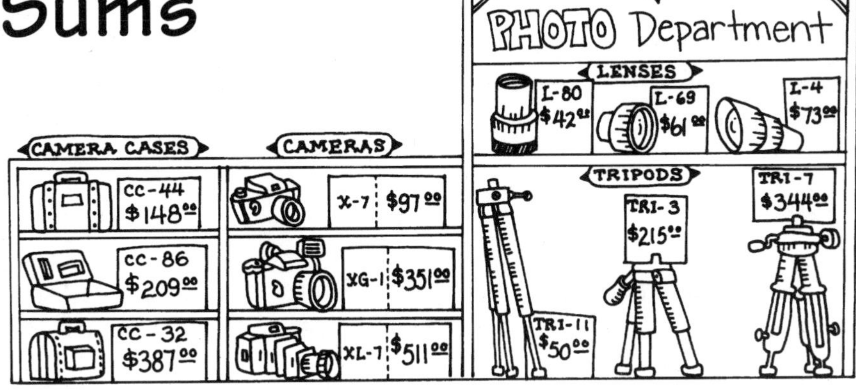

Write the estimate of each person's shopping list. Write the sum of their choices. Write the actual cost of their lists.

Raoul	*Model*	*Estimate*	*Actual*
	XG-1	______	______
	TRI-7	______	______
	L-69	______	______
	L-4	______	______
	CC-32	______	______
Sum		______	______

John	*Model*	*Estimate*	*Actual*
	X-7	______	______
	TRI-7	______	______
	L-69	______	______
	L-4	______	______
	CC-86	______	______
Sum		______	______

Mabel	*Model*	*Estimate*	*Actual*
	XL-7	______	______
	TRI-7	______	______
	L-4	______	______
	L-69	______	______
	CC-44	______	______
Sum		______	______

Cherise	*Model*	*Estimate*	*Actual*
	XL-7	______	______
	TRI-3	______	______
	L-80	______	______
	L-69	______	______
	CC-44	______	______
Sum		______	______

Who won the camera equipment? ______________________

Name ______________________ Date ______________________

NUMBER

Subtraction

Ms. Salmon has opened her own service station where she sells gas at a **profit**. *Profit* is the money she has left after subtracting the price she paid for the gas from the price she sells it for.

Solve.

1. Gas costs $0.80 a gallon. Ms. Salmon charges $1.19 a gallon. How much profit does she make per gallon?

____

2. In one month, Ms. Salmon sold $3,578 worth of fuel at a profit of $1,020. How much did the fuel cost her?

____

3. The service station made $14,505 profit the first year it opened. Last year, the profit was only $3,843. How much less money was made last year?

____

4. When the station went self-service, Ms. Salmon saved $7,548 on salaries, but spent $10,499 on new equipment. How much money did she lose?

____

5. Ms. Salmon made a profit of $2,450 selling antifreeze. She sold 2,000 cans at a total price of $10,560. How much did the cans cost her?

____

6. The station sold 6,444 cans of oil in a 2-year period. It sold 2,193 cans the second year. How many cans did it sell the first year?

____

Imagine your own store sells miniature cars.

Solve.

7. If your stock of miniature cars cost $10,500, and you sold it for $23,300, how much profit would you make?

____

8. You decide to sell the cars which cost $10,500 for $11,650. How much profit would you make?

____

9. You plan a giant sale and spend $1,050 on posters. You hire an extra salesperson for $200 and give away $40 in prizes. When figuring your profit, how much must you subtract for the cost of running your sale?

____

10. A collector of rare miniature cars offers you $307,275 for all your cars. Last year, another collector offered you $267,933 for the same collection of cars. How much more money is your collection worth this year?

____

Name ______________________ Date ______________________

NUMBER

Ordering Decimals

Trace a path from start to end. Always move to a larger decimal number.

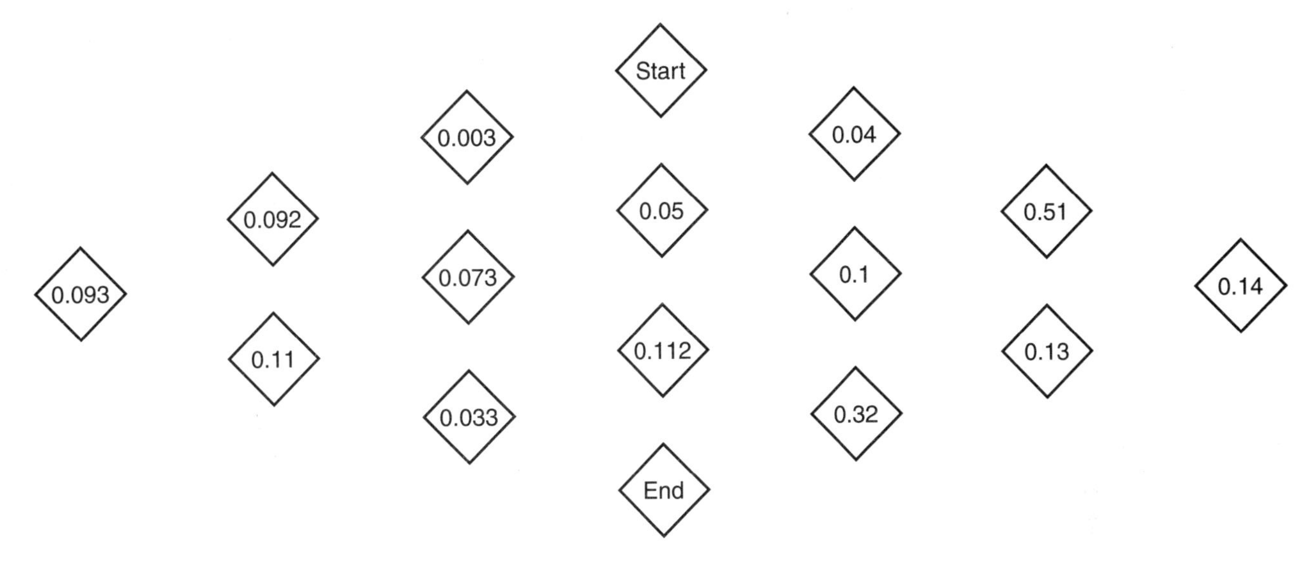

Write the numbers in order for each path.

1. Find a path six moves long.

____

____

____

2. Find a path eight moves long.

____

____

____

3. Find the four paths that are five moves long.

____

____

____

____

____

Name ________________________________ Date ________________________________

NUMBER

Complements

You can use a shortcut to solve multiplication problems. When multiplying two 2-digit numbers, you use the *complement* of the numbers. The ***complement*** of a number is the number required to bring it up to 100. To find the complement of a number, you subtract the number from 100. Try the example 93 × 95.

1. Find the complements.
100 - 93 = ______ 100 - 95 = _______

2. Multiply the complements of the numbers.
5 × 7=_______ So, ______will be in the tens and ________ in the ones places of your answer.

3. Then subtract each complement from the other number.
93 - 5 = ________

95 - 7 = ________

The answer is _______.

The number 8 will go in the thousands and the hundreds places of your answer. Sometimes the product of the complements is greater than 100. Solve 94 × 83 using complements. The complement of 94 is 6. The complement of 83 is 17.

4. First, multiply the complements of the numbers.
6 × 17 = _______ The digits in the tens and the ones places are the last two places of your answer. You have 1 in the hundreds place.

Next, subtract one complement from the other number.

94 - ________ = ________
83 - ________ = ________
These digits are in the thousands and the hundreds places of your answer.

You need to add the 1 in the hundreds place from above to the hundreds place here.
______ +1= ______ The answer is _______.

Find the product using the complements of the factors. Show your work.

5. 74 × 97 = ______________ **6.** 89 × 98 = ______________

7. 96 × 83 = ______________ **8.** 86 × 92 = ______________

Name ______________________ Date ______________________

NUMBER

Roman Numerals

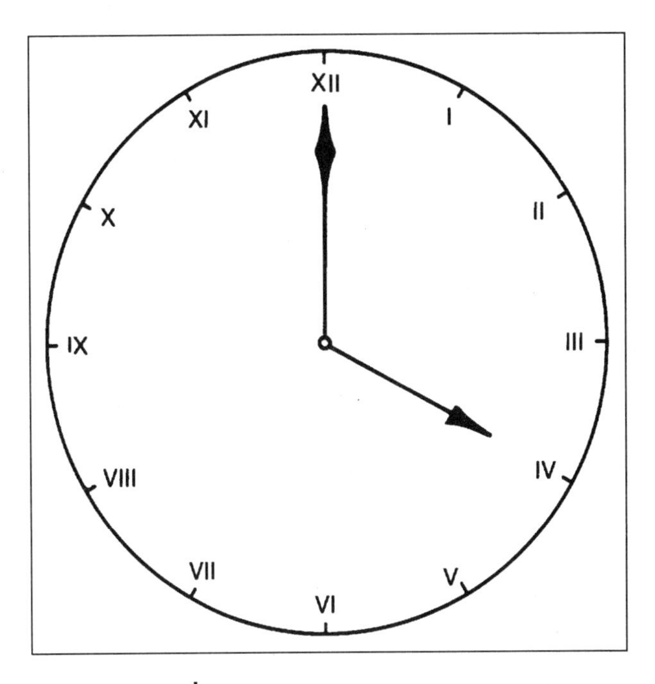

The numbers on this clock face come from the system of numbers invented by the ancient Romans. They are called **Roman numerals**. You need only three symbols (I, V, and X) to write any number from one to 39.

Use the clock to write the answers.

1. The symbol I stands for 1. The symbols V and X stand for________.

2. Grouping I's together makes the numbers 2 and 3 (II and III). The Roman numeral for 4 is IV. The Roman numeral for 6 is VI. The Roman numerals for 7 and 12 are ______________.

You can also write larger numbers by combining the three symbols I, V, and X.

Write the Roman numeral.

3. 13 = 10 + 3 = ____________________

4. 20 = 10 + 10 = ____________________

5. 16 = 10 + 5 + 1 = ____________________

6. 11 = 10 + 1 = ____________________

7. XXII + XV = ____________________

8. XXX - V = ____________________

9. XXI - IX = ____________________

10. XIX + IX = ____________________

Use a Roman numeral to write the answer.

11. How many days are there in 4 weeks? __________

12. How many letters are there in this sentence? __________

Other symbols are used to write numbers greater than 39. The symbol L stands for 50; the symbol C stands for 100.

Write the Roman numeral.

13. 67 = 50 + 10 + 5 + 2 = __________

14. 78 = 50 + 20 + 5 + 3 = __________

15. 141 = 100 + 50 - 10 + 1 = __________

16. 98 = 100 - 10 + 5 + 3 = __________

Name ______________________________ Date ______________________________

NUMBER

Magic Squares and Magic Hexagons

In a **magic square**, you find the same sum when you add the numbers of any row, column, or diagonal. This is called the **magic sum**. Look at Magic Square A.

Magic Square A

8	1	6
3	5	7
4	9	2

1. What is the magic sum of Magic Square A? ____________

You can create a new magic square from a magic square that already exists. Just multiply every number in the old square by the same factor. In Magic Square B, every number in Magic Square A has been multiplied by 4.5.

Magic Square B

36	4.5	27
13.5	22.5	31.5
18	40.5	9

2. What is the magic sum of Magic Square B? ____________

3. What is the product of 15 x 4.5? ____________

Now look at Magic Square C.

Magic Square C

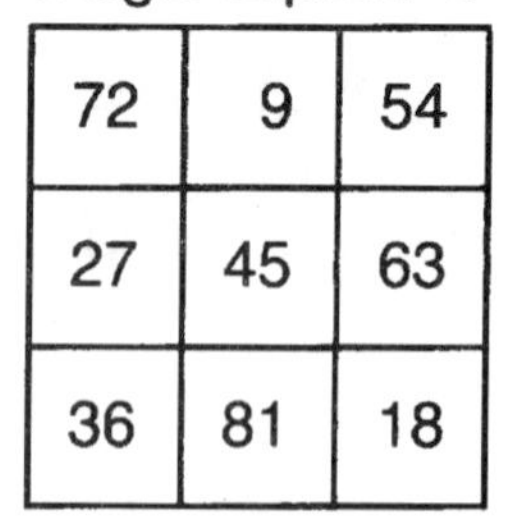

72	9	54
27	45	63
36	81	18

4. What is the magic sum of Magic Square C? ____________

5. By what factor were the numbers in Magic Square B multiplied to make Magic Square C? ____________

6. What number would you multiply Magic Square A by to create Magic Square C? ____________

Magic hexagons also have a magic sum. You find it by adding each row of boxes joined by a flat side. Complete each magic hexagon and find the magic sum for each.

7. Magic Hexagon A

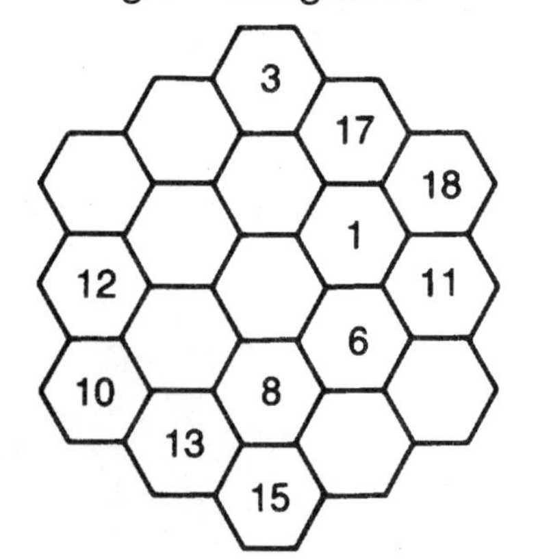

8. Magic Hexagon B

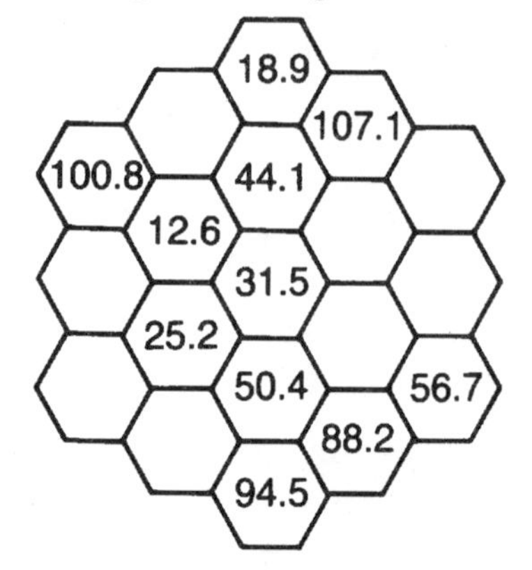

9. Magic Sum A = ____________

10. Magic Sum B = ____________

Name ______________________ Date ______________________

NUMBER

Multiplication Charts

An easy way to multiply large numbers is to break them down into smaller pieces before multiplying. For example, to multiply 37 x 134, write each factor in expanded form.

$(30 + 7) \times (100 + 30 + 4)$

Now make a chart for multiplying.

1. Fill in the chart.

Next, you add each row across to arrive at your *partial products*. List these sums to the right of your table, and add them up to find your *final product.*

x	100	30	4	
30	3,000	900	120	______
7	700	210	28	+ ______
				______

Complete the multiplication chart for each problem.

2. 325 × 25 = ■

x

+ ______

3. 275 × 14 = ■

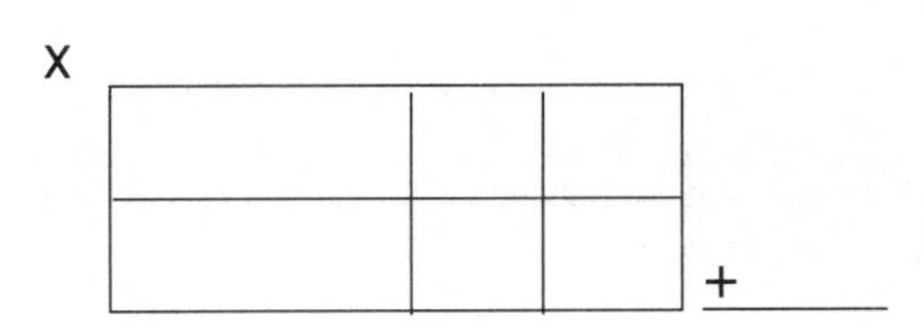

4. 173 × 91 = ■

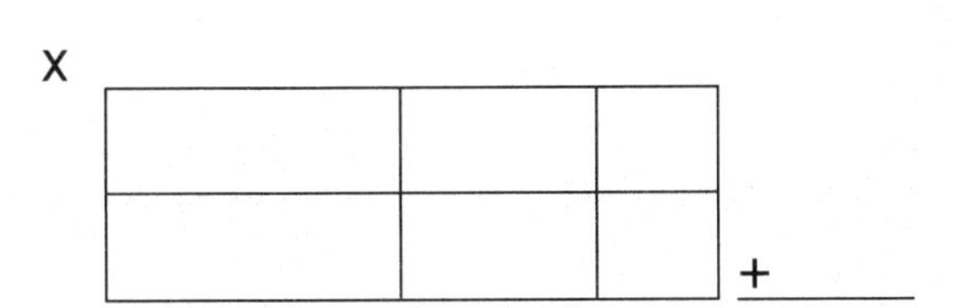

5. 3,929 × 12 = ■

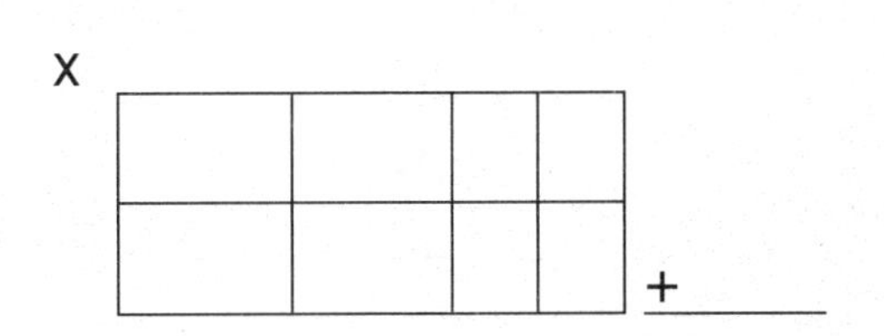

6. 999 × 4,162 = ■

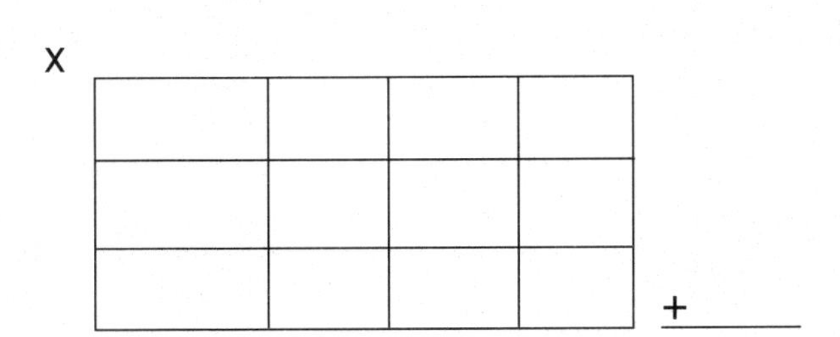

7. 330 × 2,207 = ■

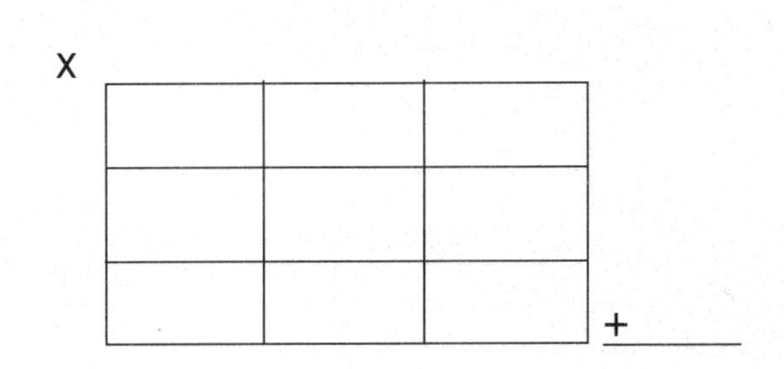

Name ______________________ Date ______________________

Dividing Decimals by Whole Numbers

Francie and Phil's Irish setter, Lacey, chases a squirrel into the North Gate of City Park. Help them follow his path out of the park through the South Gate. Connect the division problems with the correct quotients. You can't cross through any problems with your connecting line.

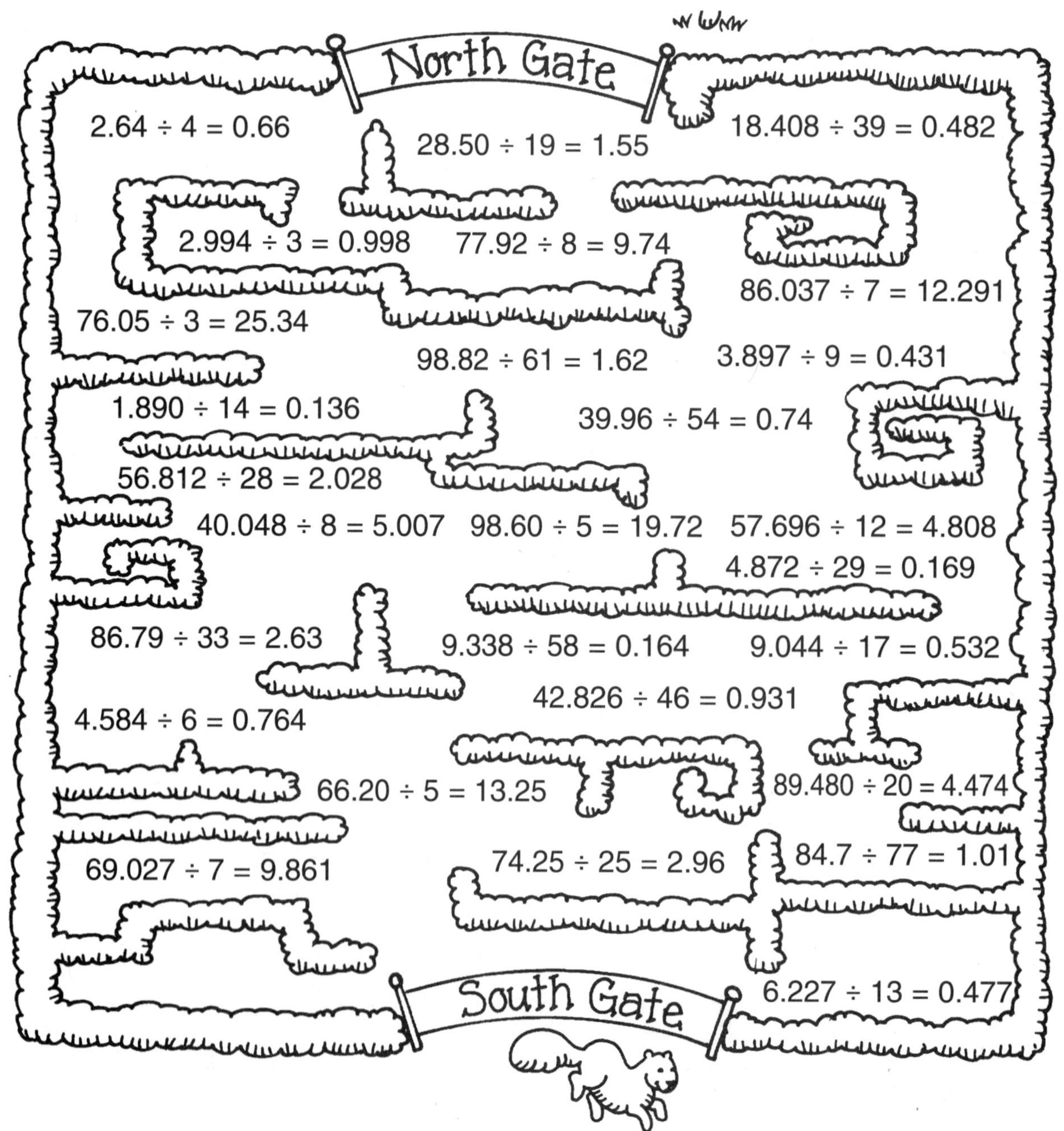

Name ______________________ Date ______________________

NUMBER

Greatest Common Factor

You can use division to find the **greatest common factor**, or **GCF**, of two numbers. Follow these steps to find the GCF of 84 and 108.

Divide the larger number by the smaller number.

$$\begin{array}{r} 1 \\ 84\overline{)108} \\ \underline{84} \\ 24 \end{array}$$

Take the divisor and divide it by the remainder.

$$\begin{array}{r} 3 \\ 24\overline{)84} \\ \underline{72} \\ 12 \end{array}$$

Repeat this step until the remainder is 0.

$$\begin{array}{r} 2 \\ 12\overline{)24} \\ \underline{24} \\ 0 \end{array}$$

The divisor that yields a remainder of 0 is the GCF of the two numbers. The GCF of 84 and 108 is 12.

Use division to find the GCF of the following pairs of numbers. Show your work.

1. 15; 6 __________

2. 85; 51 __________

3. 76; 38 __________

4. 75; 60 __________

5. 32; 80 __________

6. 99; 66 __________

7. 65; 26 __________

8. 81; 54 __________

Name ______________________ Date ______________________

NUMBER

Prime Number

Christian Goldbach was a German mathematician. He believed that every even number greater than 4 could be written as the sum of two prime numbers. A **prime number** is a whole number that has exactly two factors, itself and 1. Computers have shown this to be true for all of the even numbers up to 10,000. But it has not been proven for all even numbers. Because of this, this rule is known as **Goldbach's Conjecture**.

Write a pair of prime numbers that equal each sum.

1. Even number = Prime + Prime

Even number	=	Prime	+	Prime
6	=	3	+	
8	=		+	5
14	=	3	+	
24	=		+	

Write two pairs of prime numbers that equal each sum.

2. Even number = Prime + Prime

Even number	=	Prime	+	Prime
18	=	7	+	
36	=		+	31
44	=		+	
22	=		+	

Prime + Prime = Even number

Prime	+	Prime	=	Even number
13	+		=	18
	+		=	36
	+		=	44
	+		=	22

In 1937, a man named Vinogrador proved that any odd number greater than 5 can be written as the sum of three prime numbers.

Complete the table by writing the missing prime numbers.

3. Odd number = Prime + Prime + Prime

Odd number	=	Prime	+	Prime	+	Prime
27	=	7	+		+	3
53	=	11	+		+	
75	=		+		+	

Prime + Prime + Prime = Odd Number

Prime	+	Prime	+	Prime	=	Odd Number
7	+	7	+	13	=	27
5	+		+		=	53
	+		+		=	75

Name ______________________ Date ______________________

NUMBER

Multiplying Fractions and Whole Numbers

The pilot of a small jet is planning how to load the cargo into his jet. If the cargo is not balanced properly, the plane will not be able to fly. The pilot must arrange 1,284 pounds of cargo in the jet's three cargo compartments. Compartment 1 has a load limit of 400 pounds. Compartments 2 and 3 have load limits of 600 pounds. The pilot made a weight table, which he called Table A. It did not work because Compartment 1 would have been overloaded.

Table A
$\frac{1}{3} \times 1{,}284 = 428$ pounds in Compartment 1
$\frac{1}{3} \times 1{,}284 = 428$ pounds in Compartment 2
Place the rest in Compartment 3.

Complete the four weight tables.

Table B
$\frac{1}{4} \times 1{,}284 = 321$ pounds in Compartment 1
$\frac{1}{4} \times 1{,}284 = 321$ pounds in Compartment 2
$\frac{1}{2} \times 1{,}284 =$ ____ pounds in Compartment 3

Table C
$\frac{1}{5} \times 1{,}284 =$ ____ pounds in Compartment 1
$\frac{2}{5} \times 1{,}284 =$ ____ pounds in Compartment 2
Load the rest in Compartment 3.

Table D
$\frac{3}{8} \times 1{,}284 =$ ____ pounds in Compartment 1
$\frac{1}{4} \times 1{,}284 = 321$ pounds in Compartment 2
Load the rest in Compartment 3.

Table E
$\frac{5}{12} \times 1{,}284 =$ ____ pounds in Compartment 1
$\frac{1}{3} \times 1{,}284 = 428$ pounds in Compartment 2
Load the rest in Compartment 3.

Which table shows plans that will balance the plane's cargo? ____________

Mike has to plan how 2,655 pounds of cargo are to be loaded into a plane with five cargo compartments. Compartments 1 and 2 cannot hold more than 500 pounds each. The other three compartments have a load limit of 700 pounds. Write a way of dividing up the cargo so each compartment holds the proper amount.

Comp 1: ______________________

____

Comp 2: ______________________

____

Comp 3: ______________________

____

Comp 4: ______________________

____

Comp 5: ______________________

Date ______________________

ıals on a Number Line

The distance between the sun and the planets is enormous. You can use a number line to show how far the planets are from the sun and from one another.

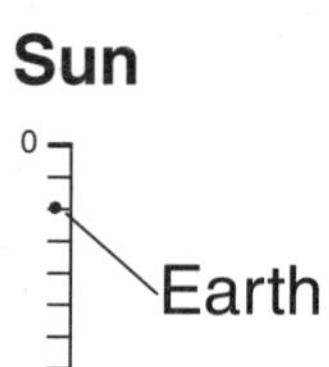

1. On this number line, the distance between numbers is divided into equal parts. How many parts are there between any two numbers?

____

2. Look at the distance between 0 and 1. Earth's distance from the sun is shown by the dot on the second point on the line. Which decimal number would show the distance between 0 and the second point on the line?

____

3. Look at the distance between 6 and 7. Neptune's distance from the sun is shown by the dot on the second point under 6. Which decimal number would show Neptune's distance from the sun?

____

4. Between which two decimal points on the line would you put 0.16? Between which two decimal points would you put 0.32?

____

5. The chart shows the location of the sun's other planets on the number line. For each location, put a dot on the number line and write the name of the planet beside the dot.

Mercury—0.08	Venus—0.16	Earth—0.20
Mars—0.32	Jupiter—1.08	Saturn—2.00
Uranus—4.00	Neptune—6.20	Pluto—8.20

6. How great is the distance between the sun and Uranus compared to the distance between the sun and Saturn?

____

Name ____________________ Date ____________________

NUMBER

Percentage

Your dog does not seem to be as energetic as usual. You decide to buy him a more nutritious dog food. In the grocery store, you compare his usual brand, Graymott's Dog Food, with Delpha Dog Food, a brand recommended by a friend.

Complete the labels.

GRAYMOTT'S DOG FOOD

Total: 150 grams/serving		
	Percentage	**Serving**
Protein	15%	22.5 g
Fat	38%	57 g
Meat by-product		28.5 g
Vitamin A		12 g
Water	9%	
Preservatives	11%	16.5 g

DELPHA DOG FOOD

Total: 150 grams/serving		
	Percentage	**Serving**
Protein	13%	19.5 g
Fat	43%	
Iron	14%	21 g
Vitamin A	13%	19.5 g
Calcium		15 g
Preservatives	7%	

You decide that Delpha Dog Food is the better brand; so, you buy a can. Delpha Dog Food has increased your dog's activity and his appetite.

Now you feed him $1\frac{1}{2}$ cans of dog food. How many grams of Vitamin A is he eating? how many grams of calcium?

Vitamin A: __________
Calcium: __________

Your bird looks tired. You go to the pet shop and the storekeeper recommends O'Connor's Bird Seed. The box has 25 servings. Your bird usually eats $2\frac{1}{2}$ servings. On special occasions, he will eat 4 servings. Complete the chart by finding how many grams of each ingredient he will eat in $2\frac{1}{2}$ servings and 4 servings.

O'Connor's Bird Seed

Total: 500 grams per box			
	Percentage	**$2\frac{1}{2}$ servings**	**4 servings**
Crushed peanuts	11%		
Sunflower seeds	8%		
Sesame seeds	39%		
Caraway seeds	42%		

Name ______________________ Date ______________________

PROBLEM SOLVING

Unit 2: Choosing the Operation

Indra and Chris are playing a board game called Philanthropy. The winner is the one who gives all of his or her money away first. Each person starts with $10,000. Decide whether you need to add or subtract, and then solve.

1. Chris lands on a space that reads, "Give up $1,500." How much money does Chris have left?

____

2. Indra has to pick a card. The card reads, "You receive $1,200." How much does Indra have now?

____

3. Chris must give Indra $2,700. How much does Chris have left?

____

4. How much does Indra have after receiving Chris's $2,700?

____

5. Indra lands on a space that reads, "Pick two cards." The first card reads "Give your opponent $4,100." How much money does Indra have left?

____

6. How much money does Chris have after taking Indra's $4,100?

____

7. Indra's second card is a secret. On Chris's turn $3,300 is given away. How much does Chris have left?

____

8. Indra's space reads, "Lose the same amount your opponent last lost." How much does Indra have left?

____

9. Chris loses $2,800. How much does Chris have left?

____

10. Indra also loses $2,800. How much does Indra have left?

____

11. Chris must give $3,700 to Indra. But Indra's secret card reads, "Give your opponent the amount of money he or she tries to give you instead." How much does Chris have now?

____

12. How much does Indra have?

____

Name ______________________ Date ______________________

PROBLEM SOLVING

Using Broken-Line Graphs and Bar Graphs

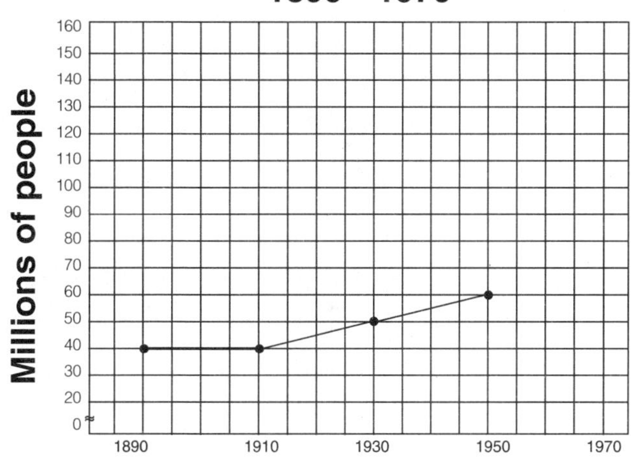

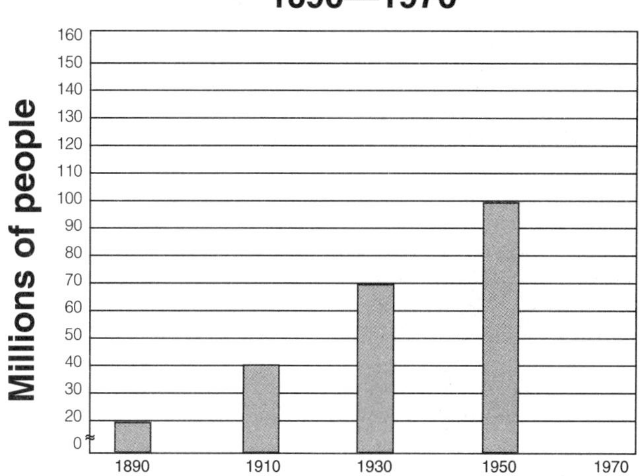

These graphs show the urban and rural populations of the United States during a period of time. Use the graphs to complete this page.

1. The rural population in 1970 was 10 million less than it was in 1950. Write this information on the line graph. Continue the line.

2. The urban population in 1970 was 50 million more than it was in 1950. Write this information on the bar graph.

3. In 1930, about how many more Americans lived in urban areas than in rural areas?

____

4. After 1930, which part of the population grew more rapidly?

____

5. Between which years does the graph show that the rural population is larger than the urban population?

____

6. In what year were the rural and the urban populations the same?

____

7. If the urban growth trend continues, would the urban population of 1980 fit on this bar graph?

____

8. Since suburbs and cities are related, what can one guess about the growth of suburbs since 1950?

____

Name ______________________ Date ______________________

PROBLEM SOLVING

Writing a Number Sentence

The Drones and UBX-42 are two new music groups whose albums are competing for first place on the music-popularity charts.

Write and solve by using number sentences that describe the accomplishments of both bands.

1. In the recording studio, the Drones taped a song that lasted 48.7 seconds. They were not satisfied with it, so they taped it again. The second taping lasted 2.9 seconds less than the first. The third taping lasted 0.47 seconds more than the second. What number sentence expresses the length of the third taping? How long was the third taping? Altogether, how long were the three recordings?

____

____

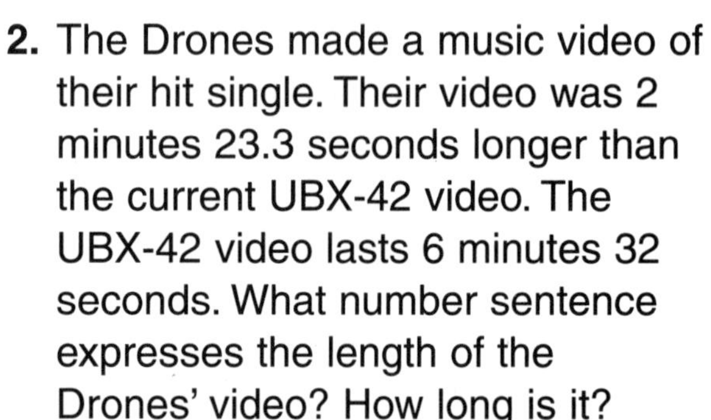

2. The Drones made a music video of their hit single. Their video was 2 minutes 23.3 seconds longer than the current UBX-42 video. The UBX-42 video lasts 6 minutes 32 seconds. What number sentence expresses the length of the Drones' video? How long is it?

____

____

3. The Drones' album has sold 1.35 million copies in 8 months. Their manager said he expects the album to sell another 0.42 million in the next 2 months. By the end of the year, he thinks the album will have sold 0.74 million more copies than in the first 10 months. What number sentence shows the amount of sales expected for 12 months? How many copies might be sold?

____

____

4. On the first night of a 3-concert tour, the Drones drew 47,320 fans. The second concert, 2,420 fewer fans attended than the first concert. The third concert, 1,755 more fans attended than the second concert. What number sentence expresses the number of fans who attended the third concert? How many fans were at the third concert?

____

____

Name ______________________ Date ______________________

PROBLEM SOLVING

Estimation

You are the set designer for a play. Below is a list of prices of the supplies that you need.

paint	$9.76 per gallon	wood	$1.68 per foot
fabric	$4.15 per yard	nails	$1.89 per box
canvas	$30.45 per roll	rope	$2.30 per yard
brushes	$5.20 each	masking tape	$0.90 per roll

Use estimation or find the exact answer as the problem requires.

1. A lumberyard has offered to supply $200 worth of free merchandise. Your design calls for 100 feet of wood and 12 boxes of nails. Could you get 5 more feet of wood without having to pay the lumberyard any money?

____

2. You need 76 yards of fabric and 9 rolls of canvas. The store gives a discount on any order that totals more than $600. How many more rolls of canvas would you have to buy to qualify for the discount?

____

3. You have brought $300 to the hardware store. You want to buy 21 gallons of paint, 6 brushes, and 12 rolls of masking tape. Do you have enough money to buy 30 yards of rope as well?

____

4. You wrote checks for 5 yards of fabric to Cloth Cutters, Inc., and for 10 yards of rope to Ward Hardware. You forgot to fill in the names. Which store should have received the check for the larger amount?

____

5. A friend has asked you to order some supplies for him. He wants 2 gallons of paint, one brush, and 54 feet of wood. He wants to spend about the same amount on canvas as he spends on the above items. How many rolls should you order for him?

____

6. You need 4 new curtains. Each will use 12 yards of fabric, and each will need 10 yards of rope. Your budget is $275. Will this be enough?

____

Name ______________________ Date ______________________

PROBLEM SOLVING

Solving Two-Step Problems/ Making a Plan

Two families take a long-distance hike from the town of Redbird to Sprucewood State Park. Each family takes a different route. On normal ground, the hikers cover 10 miles per day. In the mountains, they travel about 7 miles per day. The map shows the distance and terrain of each route.

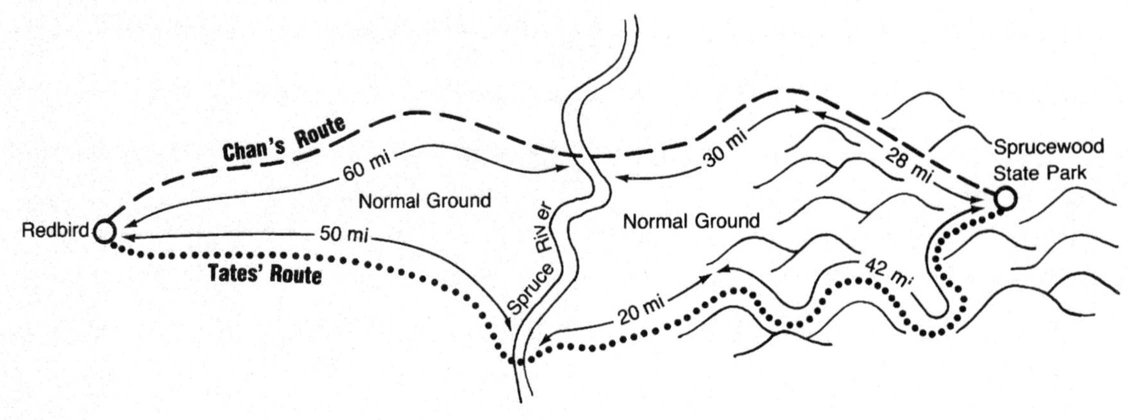

Complete the plan by writing in the missing step.

1. The Chans walk over normal ground for 4 days. How many more miles must they walk to reach the river?

Step 1: ______________________

____

Step 2: Subtract to find how many miles are left before they reach the river.

2. On the Tates' route, how many more miles is it from the river to the park than it is from Redbird to the river?

Step 1: Add to find how many miles it is from the river to the park.

Step 2: ______________________

____

Make a plan and solve.

3. A shortcut could get the Chans to the river in just 4 days. Walking at their usual rate, how many miles would they save?

____

4. One family has walked through mountains for 4 days. They are now 14 miles from the park. Which family is it?

____

5. The Tates have traveled for 7 days. Have they reached the mountains yet?

____

6. According to the map, how many miles longer is the Chans' route than the Tates' route?

____

Name ______________________________ Date ______________________________

PROBLEM SOLVING

Writing a Number Sentence

When the Sanders bought a new motorboat, they started keeping a ship's log. One day while in heavy surf, water splashed on the pages and the ink ran. Write a number sentence in the space provided that will help you to fill in each blank. Then complete the ship's log.

Number Sentences

We took our new boat on our first overnight fishing trip. We motored 45 miles Saturday. We motored (1) __________ miles Sunday, and 102 miles in all. Scott took the boat to the marina today and spent $26 on gasoline and oil. The gasoline cost $18, and the oil cost (2) $ __________.

1. ____________________

2. ____________________

Hank likes to fish. He just bought 4 sets of deep-sea fishing tackle for $52 each, or (3) $ __________ altogether. Scott is trying to economize. Instead of buying new ski ropes, he made his own. He bought 225 feet of rope and cut it into 3 equal pieces. Now we have new ropes, each one (4) __________ feet long.

3. ____________________

4. ____________________

We intend to take our time sailing to New Orleans. We have 81 miles to travel and 9 days to cover the distance. We only need to travel (5) __________ miles per day.

5. ____________________

Today, Hank clocked the boat's speed. Our cruising speed is 11.2 miles per hour. In 5 hours, we could travel (6) __________ miles. When we got back to the dock, it was time for chores. Everybody has an assigned job. Kerri spent 3 hours on her jobs. It took her 2.4 hours to scrub the boat, and she spent (7) __________ putting the fishing gear away. Scott polished all the brasswork. He spent 1.2 hours polishing the cleats, 2.3 hours polishing the railing, and 1.5 hours on the rest. He spent (8) __________ hours in all and says that Kerri and he should switch jobs next time.

6. ____________________

7. ____________________

8. ____________________

Name ______________________ Date ______________________

PROBLEM SOLVING

Estimation

Manuela is planning dinner. She wants to be sure that each dish will be ready in time. She decides that dinner should start with soup at 5:00 P.M. Help Manuela plan her meal.

Dish	How much?	Estimated cooking time	Starting time	Estimated serving time
Soup	4 cups			
Beef	5.37 pounds			
Potatoes	4			
Beans	4 cups			
Dessert	6 baked apples			

Use the information below to fill in the table. Think about whether it would be better for Manuela to *overestimate* or *underestimate*.

1. The beef, potatoes, and lima beans make up the main course and should be served at the same time.
2. Manuela needs to estimate how long it will take to eat the soup so that she can figure out when to serve the main course.
3. She needs to estimate how long it will take to eat the main course so that she can figure out when to serve dessert.
4. She should allow 30-35 minutes of cooking time for each pound of beef.
5. Baking one potato takes 40 minutes. When baking more than one potato, allow at least 5 extra minutes for each.
6. The soup should be brought to a boil.
7. The beans are to be cooked in 2 cups of boiling water. Each cup of beans will cook for 2 1/2 minutes.
8. Manuela knows it takes 5 minutes for 2 cups of water to boil.
9. Baking the apples will take 40-45 minutes. They need to cool for at least 15 minutes.

Name ______________________ Date ______________________

PROBLEM SOLVING

Identifing Extra Information

Solve each problem. Draw a line through the information you do not need.

1. The Statue of Liberty is 45.32 m tall. Miss Liberty herself is 33.45 m tall. Her head is 5.18 m. How much height does her raised arm add to the statue? ______________

2. Each eye of the Statue of Liberty measures 0.75 m in width. The mouth is 0.80 m wide. Its head is four times as wide as each of its eyes. How wide is the head? ________

3. The granite pedestal is 26.7 m high and has columns 21.8 m high on it. It stands on a foundation that is 19.5 m high. How tall is the statue, including the pedestal and the foundation? ______________

4. The copper covering the statue is 2.3 mm thick. 90,000 kg of copper and 101,250 kg of steel were used to make the Statue of Liberty. How much more steel than copper was used? ______________

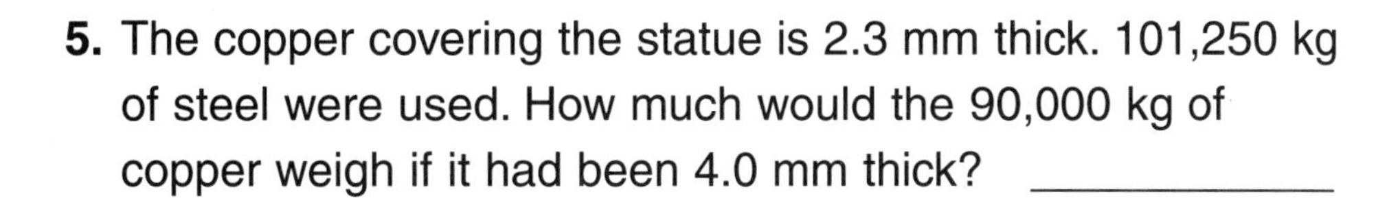

5. The copper covering the statue is 2.3 mm thick. 101,250 kg of steel were used. How much would the 90,000 kg of copper weigh if it had been 4.0 mm thick? ________

6. The statue's designer, Frederic Bartholdi, was 40 years old when he began to work on it. He finished work on it in 1884, 10 years later. He died in 1904. In what year was he born? ______________

7. Frederic Bartholdi used three models. The first was only 1.25 meters tall, the second 2.85 meters, and the third was 8.8 times the height of the first. How tall was the third model? ______________

Name ______________________ Date ______________________

PROBLEM SOLVING

Solving Multistep Problems

The sixth-grade class is hosting a costume party. There will be prizes for the most original costume, the funniest costume, and the scariest costume.

Solve.

1. For her costume, Maria decides to dress up as a potted fern. She buys a roll of foam rubber 12 meters long and 1 meter wide. It takes 2 meters of foam rubber to make one leaf. How many leaves can she make and still have 2 meters left for the base of her costume?

   ____

2. James, Elspeth, Karen, Richard, Steve, and Marvin come to the party as a crowd. Richard has to leave early, but 2 friends of Karen's join the crowd. By the end of the party the crowd is half its original size. How many people have left the crowd?

   ____

3. Mark decides to dress as his favorite food. He makes meatballs out of papier mâché and the noodles out of ribbon. The ribbon comes in spools 8 meters long. If Mark makes each noodle 2 meters long, how many spools of ribbon will he need to make 45 noodles?

   ____

4. Christina comes to the party dressed as a caterpillar, and she shares the prize for the funniest costume with Mark. Since no one dressed in a scary costume, the judges decide to combine the $10 prize for scariest costume with the $15 prize for funniest costume, and to divide it equally between Christina and Mark. How much money will each one receive?

   ____

5. Alfie and Marie are tied for the $25 prize for the most original costume. They give some of the money to Bobby, who helped them make the costumes, and divide the rest evenly. If they each keep $8.50, how much money did they give Bobby?

   ____

Name ______________________ Date ______________________

PROBLEM SOLVING

Interpreting the Quotient and the Remainder

Imagine that you are an astronaut on a mission in space. You must find out if the planet Dither can support a space colony. You must take samples of the soil and air, photograph the landscape, and conduct experiments.

Solve.

1. The soil tester deposits 9 soil samples at a time in lab trays. There are 321 samples to be tested. How many lab trays are being used? Are all the lab trays completely full? ______________________

2. There are 42 air tanks to be taken to a testing site. If 5 air tanks fit into a tank carrier, how many full carriers will there be? How many tanks will be left over and then put into a partly full carrier? ______________________

3. The landscape camera holds enough film to take 60 pictures. One day, Mission Control sends you out to take pictures in groups of 7. How many groups of 7 pictures will you be able to take? How many pictures could you take with the film that is left over? The next day, Mission Control sends out 4 astronauts to take 60 more pictures. How many pictures can each astronaut take? ______________________

4. You and the 3 other members of your crew must conduct 22 experiments in the 7 days you are going to be on the planet. How many experiments is that per person? how many per day? If one of the leftover experiments was given to each person, how many people would have to do an extra experiment? How many experiments would be done on the last day? ______________________

5. Your space car can carry 8 pounds of rock samples on each trip. If you have 68 pounds of samples to take back to the ship, how many trips must you make in the space car? ______________________

6. It takes 4 fuel packs to get the space car to and from Testing Site X. If you have 35 fuel packs, how many times can you go to and from the testing site? How many fuel packs will be left over? ______________________

Name ______________________ Date ______________________

PROBLEM SOLVING

Using a Time-Zone Map

TIME ZONES OF THE CONTIGUOUS UNITED STATES

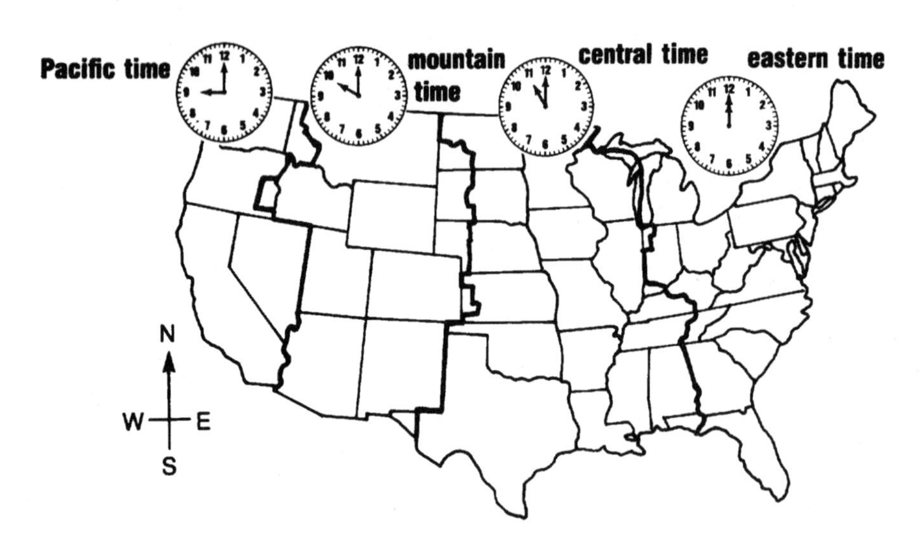

Name that time zone!

1. Which time zone is this state in?

____

2. Which time zone is this state in?

____

3. If it is noon in this state, what time is it in most of this state?

____

4. Which two time zones is this state in?

____

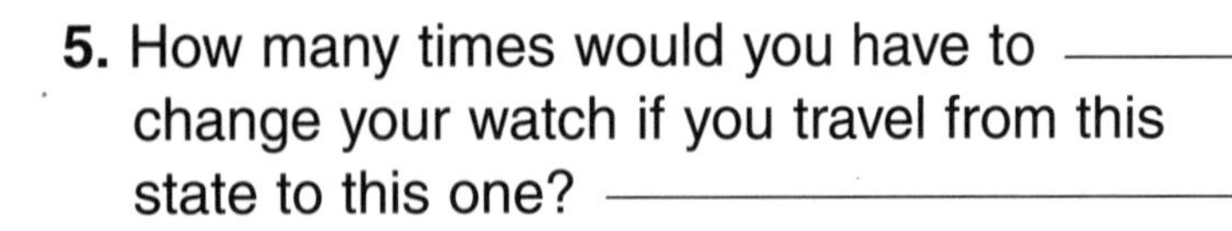

5. How many times would you have to change your watch if you travel from this state to this one?

____

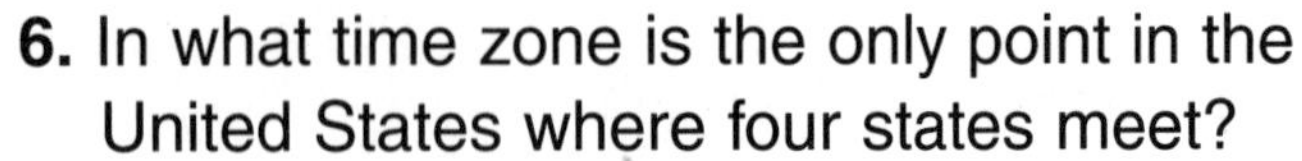

6. In what time zone is the only point in the United States where four states meet?

____

Name ____________________ Date ____________________

PROBLEM SOLVING

Using a Circle Graph

Use the circle graphs to solve.

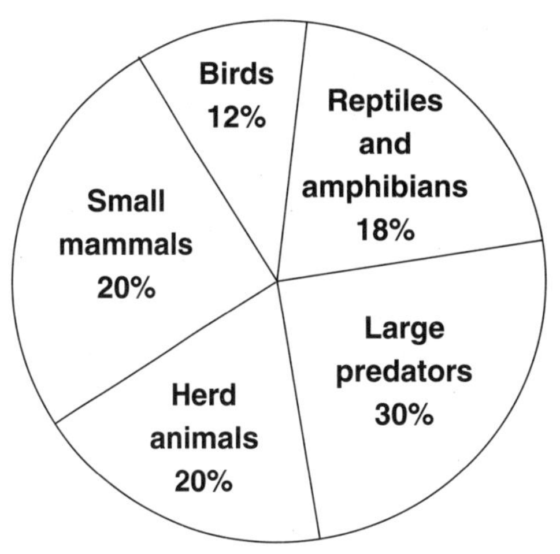

ANIMALS AT THE COLORADO CANYON INSTITUTE (700 animals in all)

1. How many reptiles and amphibians are there at the zoo?

____

2. Of the herd animals at the zoo, 36 are gazelles. How many of the herd animals are not gazelles? How many more herd animals are there than reptiles and amphibians?

____

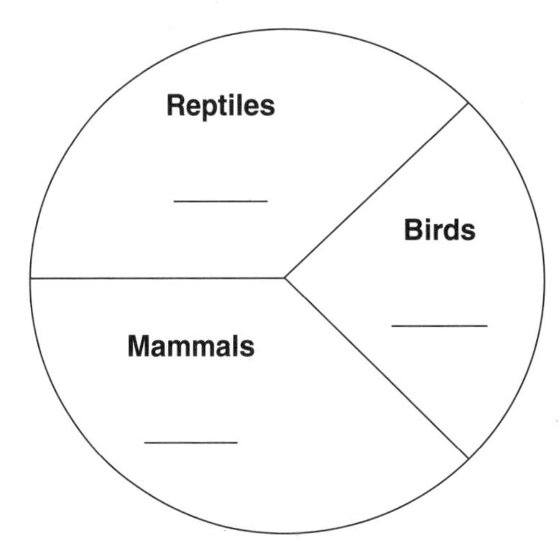

ENDANGERED SPECIES (140 animals in all)

3. There are 28 birds in the endangered species section at the zoo. What percent is this? Write the percent on the graph.

____

4. Of the endangered species, 40% are mammals. Of the predators, 30% are small predators. Which group has more animals, endangered mammals or small predators? how many more?

____

5. What is the percent of reptiles at the zoo? Write the percent on the graph. How many reptiles are there?

____

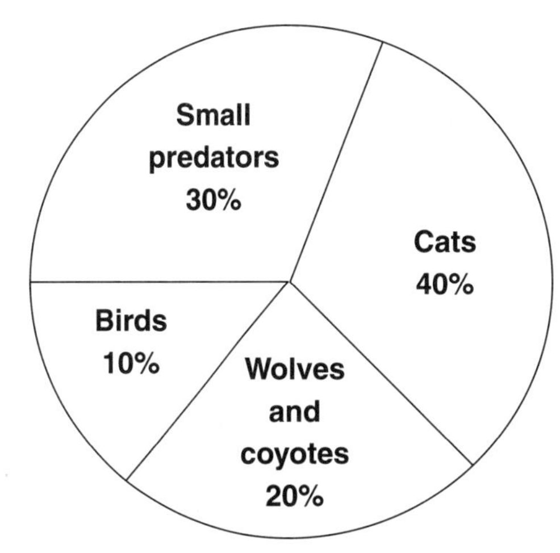

PREDATORS (210 animals in all)

6. How many wolves and coyotes are among the predators at the zoo? Is this more or less than the number of endangered birds?

____

7. Which bird group is larger, predators or endangered species? how much larger?

____

Name ______________________ Date ______________________

PROBLEM SOLVING

Interpreting a Graph

The two graphs show information about two different schools. Complete the graph on the right by using the facts below. Then answer the questions. Remember to read both graphs carefully.

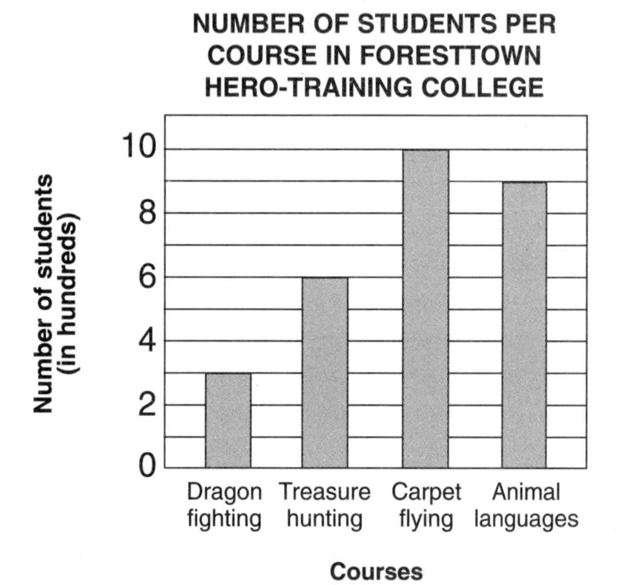

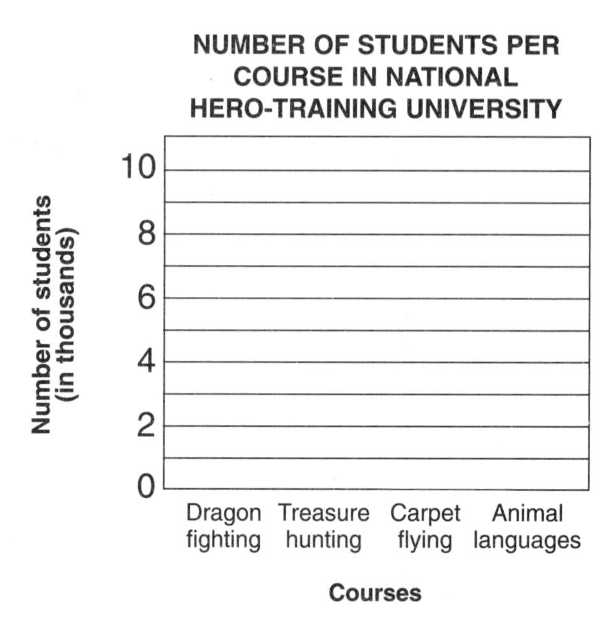

Students in each course in the university include:
Dragon fighting-2,000; Treasure hunting-6,000;
Carpet flying-1,000; Animal languages-10,000.

1. How many students are taking the animal-languages course at the college?

____

2. Which course has the same number of students in both schools?

____

3. How many more students are there in the college's animal-languages course than are in the treasure-hunting course?

____

4. How many more people are taking animal languages at the university than are taking it at the college?

____

5. Last year 150 people took treasure hunting in a correspondence course by mail. Which school has 4 times that number of students in its treasure-hunting course?

____

6. The college offers a course called Rescuing kingdoms. It has 1,000 students. If you write this information on the college's graph, would it be one of the longest bars or one of the shortest?

____

Name ______________________ Date ______________________

LOGIC

Unit 3: Money Equivalents

1. Kazuko has one dollar and thirty cents in nickels, dimes, and quarters. She has twice as many nickels as dimes and twice as many dimes as quarters. How many of each coin does she have?

____

2. Jeb has five dollars' worth of coins, but none of them are nickels or quarters. He has one half-dollar. He has a total of one hundred coins. How many dimes and pennies does he have?

____

3. Nikki has six coins. She has some dimes, nickels, and pennies. She has at least one of each type of coin. She has more nickels than dimes and more dimes than pennies. What is the total value of her coins?

____

4. Hulya cannot make change for a half-dollar, a quarter, a dime, or a nickel. What is the greatest number of coins she can have and what are they?

____

5. Ruth has twice as many dimes as nickels, and five times as many nickels as quarters. She has three dollars in coins in all. How many of each coin does she have?

____

6. Reggie has two dollars' worth of nickels, dimes, and quarters. He has a total of twenty coins. He has the same number of dimes as quarters. How many of each coin does he have?

____

7. Brad has $2.50 worth of three kinds of coins. The largest coins are worth less than the smallest ones, but their total value is twice as much as that of the smallest ones. The medium-sized coins have a total value of two of the largest coins. What are the coins and how many of each are there?

____

Name ______________________ Date ______________________

LOGIC

Venn Diagram

A **Venn diagram** is used to show how sets are related.

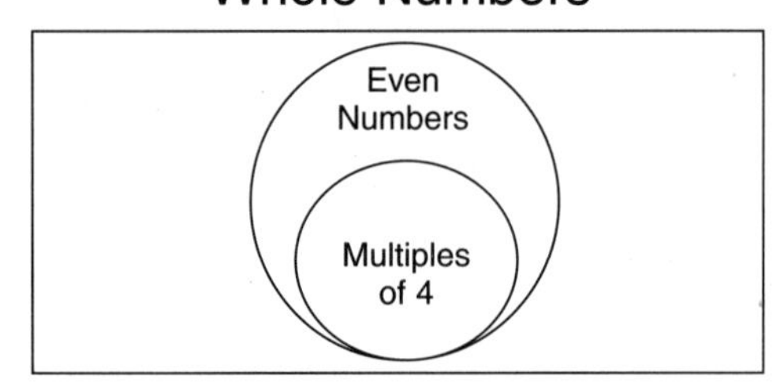

This diagram shows that:
1. All multiples of 4 are even numbers.
2. Some even numbers are not multiples of 4.

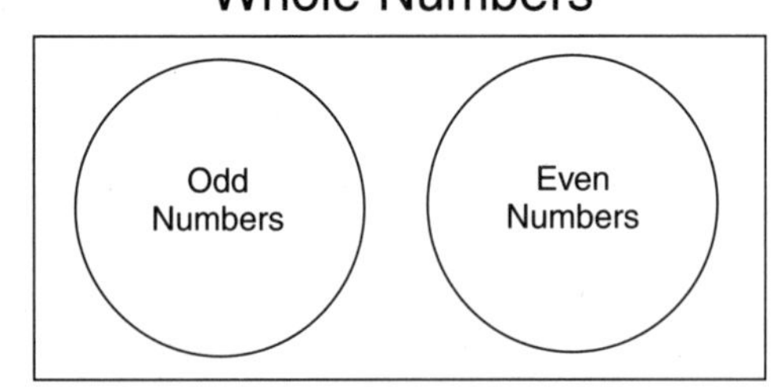

This diagram shows that:
1. No odd numbers are even numbers.
2. No even numbers are odd numbers.

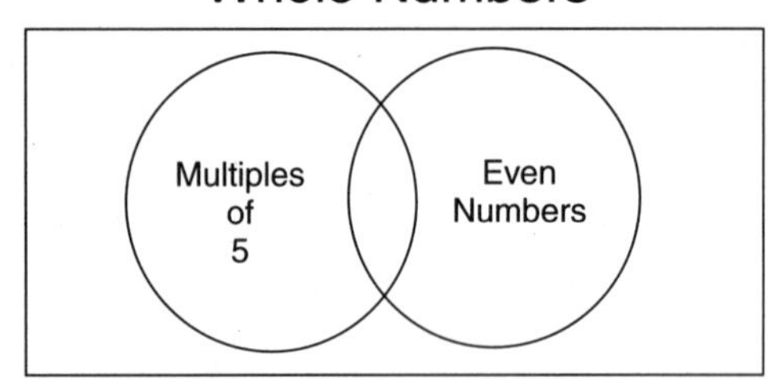

This diagram shows that:
1. Some even numbers are multiples of 5.
2. Not all even numbers are multiples of 5.
3. Some multiples of 5 are even numbers.
4. Not all multiples of 5 are even numbers.

Write two correct statements for each diagram.

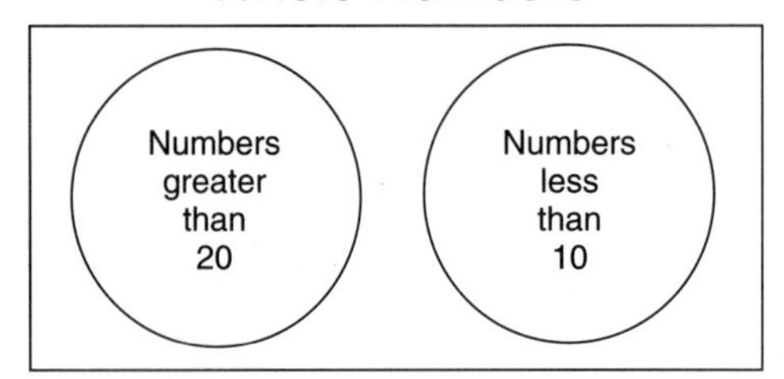

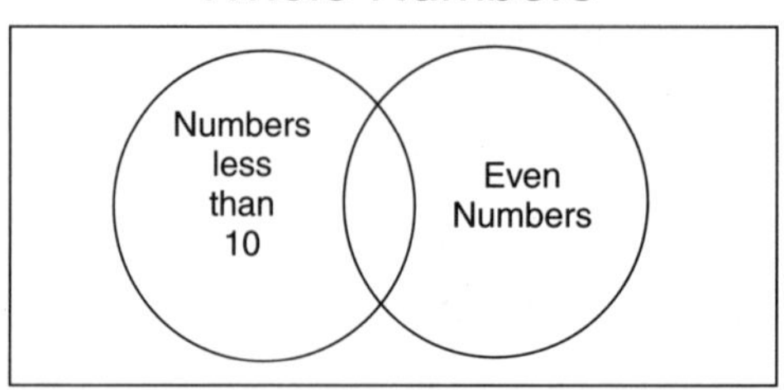

1. ______________________

2. ______________________

3. ______________________

4. ______________________

Draw a Venn diagram for this pair of statements.

5. All multiples of 10 are even numbers.
   Some even numbers are multiples of 10.

Whole Numbers

Name ______________________________ Date ______________________________

LOGIC

Winning Strategy

Everyone knows the three-in-a-row game called Tick–tack–toe. Did you know that the player who goes first can never lose and the player who goes second can never win as long as both players are smart and careful? In the games that follow, you can play both sides. In that way, you can work out a winning strategy.

1. a. Finish this game so that the first player wins. Player 1 marks an X; Player 2 marks an 0.

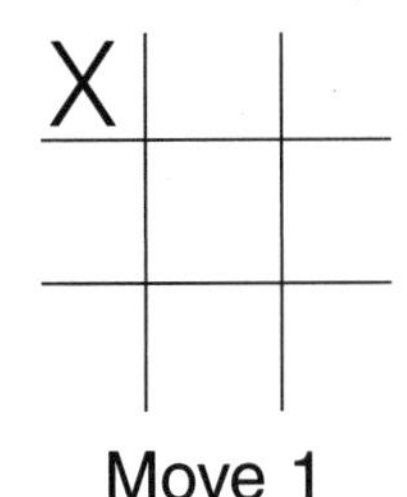

Move 1

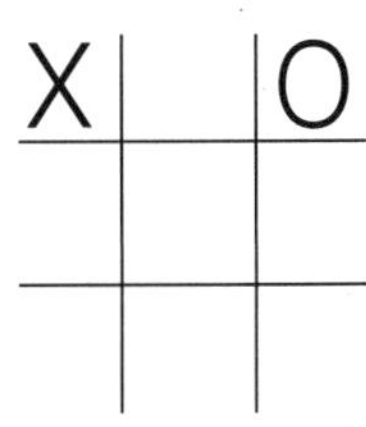

Move 2

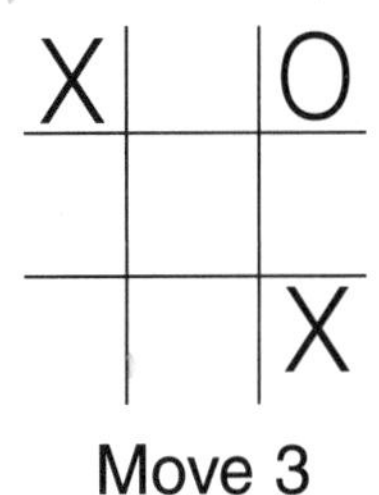

Move 3

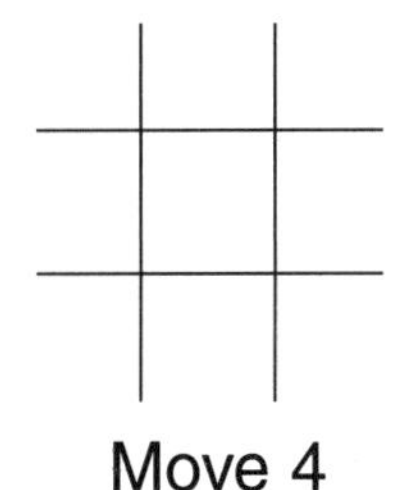

Move 4

b. *Analysis:* Player 1 has set a trap by forcing Player 2 to mark an 0 in the center. As a result, Player 1 has two different ways to get three-in-a-row.

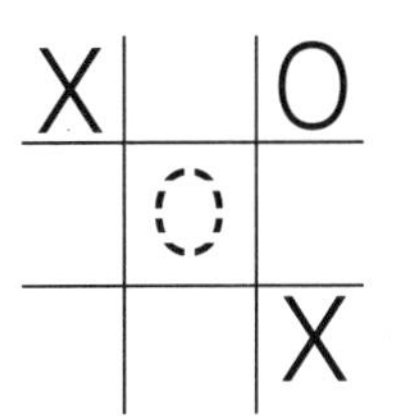

Complete each game so that the first player wins. In each game, Player 1 can set a trap by forcing Player 2 to make a certain move.

2.

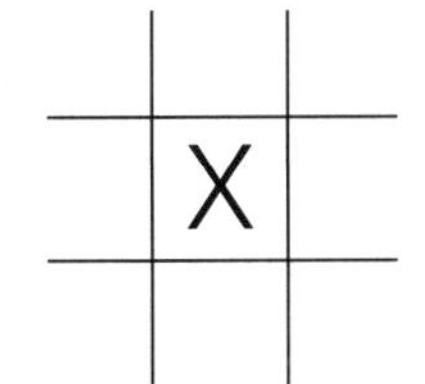

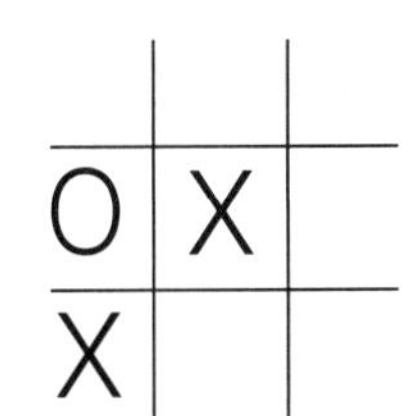

3.

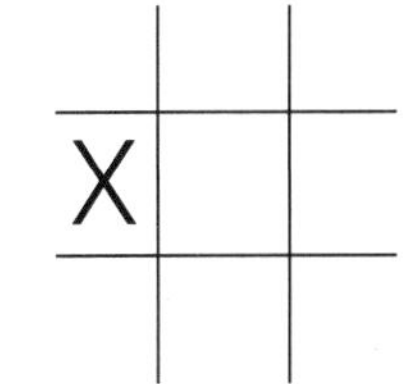

Complete each game so that Player 2 does not lose; that is, the game ends in a draw.

4.

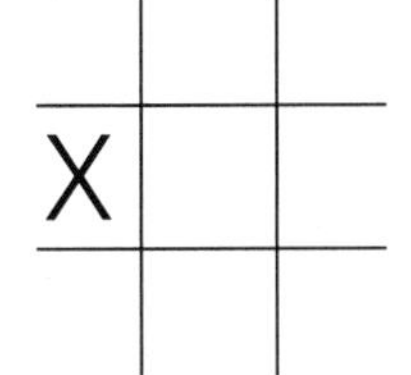

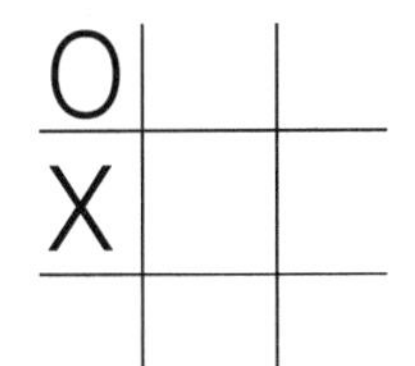

5.

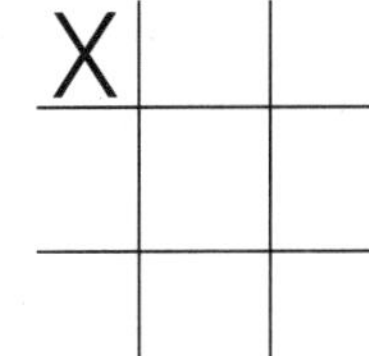

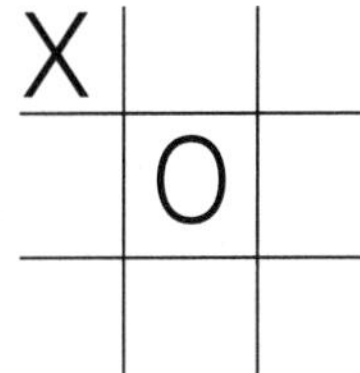

Name ______________________________ Date ______________________________

LOGIC

Logic Table

Marcus, Cindy, Eileen, Steve, Eric, and Meredith are all planning to go to a masquerade party. Steve went to a costume store and picked out six costumes: a two-person horse suit, a bunny suit, a Roman toga, and two matching pirate costumes.

a. Neither Marcus nor Cindy would wear a costume that someone else was wearing.

b. Eric, who didn't go as a pirate, wore the same costume as Steve.

c. Marcus didn't wear the toga.

d. Meredith wasn't the bunny.

Who wore each costume?

Answer each question.

1. How many correct places will there be in each row and each column of the table?

____

2. If you know that two people are wearing the same costume, which could it be?

____

3. If you know that two people are wearing costumes that no one else is wearing, which costumes could they be?

____

4. Can Steve go as a pirate to the party? Why or why not?

____

Use the table to keep track of the information in the clues.
Place an X where it is not true and an O where it is true.

	Horse	Horse	Bunny	Toga	Pirate	Pirate
Marcus						
Cindy						
Eileen						
Steve						
Eric						
Meredith						

Write the costume each person wore.

Marcus: ______________ Cindy: ______________ Eileen: ______________

Steve: ______________ Eric: ______________ Meredith: ______________

Name ______________________________ Date ______________________________

LOGIC

Completing a Diagram

Look at the diagram of the soccer-league playoffs.

BROWNSVILLE SCHOOL SOCCER PLAYOFFS

Round 1	Semifinals	Finals	Championship	Champions
Cougars	Cougars	Cougars	?	Sharks
Eagles	Mustangs	Busters	?	
?				
Rhinos	?	?		
?				
?	Hornets	Panthers		
Goal-Getters				
Giants	Sharks			
Sharks				
Kangaroos	Titans			
Titans				
?	Flyers			
Kickers				
Panthers	Panthers			
Bears				

Use logic to write the answers.

1. Which team beat the Kickers? ______________________________

2. Which team beat the Bears? ______________________________

3. Who beat the Rhinos? ______________________________

4. Which team did the Titans beat? ______________________________

5. Who beat the Eagles? ______________________________

6. Whom did the Goal-Getters play in Round 1? ______________________________

7. Whom did the Busters lose to? ______________________________

8. Whom did the Panthers lose to? ______________________________

Name ______________________________ Date ______________________________

LOGIC

Determining Age

Solve the word problems.

1. Jack is 10 years older than Martha. The sum of their ages is 5 times the difference between their ages. How old are they?

Jack: ____________ Martha: ____________

2. Sue is as old as her sisters Nancy and Jane together. Last year, Nancy was twice as old as Jane. Two years from now, Sue will be twice as old as Jane. How old are they?

Sue: ____________ Nancy: ____________ Jane: ____________

3. When Alexandra was 10 years old, Juanita was 20 years old. When Juanita was 15 years old, Carleton was 20 years old. When Carleton was 20 years old, Emilio was 25 years old. When Emilio was 10 years old, Eirendira was 5 years old. Eirendira was born in 1964. How old were the others when she had her twenty-first birthday?

Alexandra: ____________ Juanita: ____________

Carleton: ____________ Emilio: ____________

4. Brad is $\frac{3}{4}$ the age of Manuel. Manuel is twice as old as Benigno. José is 3 years and 2 months younger than Brad. Douglas is $1\frac{1}{2}$ times the age of José and a month older than Manuel. Benigno is 19 years and 4 months old. How old are the others?

Brad: ____________ Manuel: ____________

José: ____________ Douglas: ____________

5. Yukio's father is 6 times as old as his son. Yukio's mother is 5 times as old as her son. In 22 years 4 months, Yukio will be as old as his mother is now. At that time, Yukio's father will be 55 years 10 months. How old are all of them now?

Yukio: ____________ Father: ____________

Mother: ____________

Name ______________________ Date ______________________

LOGIC

Organizing Clues

To solve certain types of logic problems, you must organize a series of clues and use the clues to decide what the missing facts are. A table can help you keep track of what you have learned from the clues.

Mr. Richards, Mrs. Bellamy, Mr. Jimenez, and Ms. Eshghi are all teachers at Sunnyvale High School.

- Each subject is taught by two teachers. No one teaches more than two subjects.
- Mrs. Bellamy teaches history.
- Mr. Jimenez likes languages but does not teach them.
- Ms. Eshghi was born in the West.
- The Spanish teachers were born in New York and Vermont.
- The math teachers were born in Idaho and California.
- One of the French teachers also teaches Spanish.
- The French teachers were born in New York and Idaho.
- Mr. Richards does not teach math or history.

1. How many correct places will there be in each row and in each column of the table? ______________

2. How do you know what Mr. Jimenez teaches?

____

3. Use the table below to organize your clues. Put a large 0 in a box when you know that the column with the name of the subject is definitely taught by the person whose name begins that row. Put an X in a box when you are sure that the person does not teach that subject.

SUBJECT

	Math	**History**	**French**	**Spanish**
Mr. Richards				
Mrs. Bellamy				
Mr. Jimenez				
Ms. Eshghi				

4. Where were each of the teachers born?

Mr. Richards: ______________ Mrs. Bellamy: ______________

Mr. Jimenez: ______________ Ms. Eshghi: ______________

Name ______________________________ Date ______________________________

LOGIC

Determining Distance

At the Allentown Auto Race, all the cars broke down before the end of the race. The judges gave the first prize to the driver whose car had gone the greatest distance.

- Buzz Benson's car stopped after completing the first fifth of the last fifth of the race.
- Celia Cannon's car stopped with a third of the race left to finish.
- Donna Dorn's motor gave out after completing the first half of the thirteenth sixteenth of the race.
- Ed Einstein's car stopped halfway into the second half.
- Eileen Cowell went only half as far as Donna Dorn.
- Eleanor Wong went two thirds as far as Donna Dorn
- Manuel Ruiz stopped one ninth into the last third of the race.

Solve. Write each fraction in lowest terms.

1. How far did Buzz Benson's car go before breaking down?

____

2. Where did Celia Cannon's car stop?

____

3. How far did Donna Dorn's car go in the race?

____

4. How far did Eleanor Wong's car go in the race?

____

5. Did Manuel win the race? How far did his car go?

____

6. Whose car came in last? How far did the car go before stopping?

____

7. In what order did the drivers finish?

____

Name ______________________ Date ______________________

LOGIC

Determining Time

You must use **logic** to answer these questions. Read the clues; then fill in the table below each set of clues.

One night:

Annie went to sleep at 10:00 P.M.

Jamal went to sleep 20 minutes before Annie.

Li went to sleep 40 minutes after Jamal.

David went to sleep 50 minutes after Enid.

Enid went to sleep 20 minutes before Fred.

Fred went to sleep 35 minutes after Annie.

Jorge went to sleep 30 minutes after everyone else.

The next morning:

Annie woke up 20 minutes after Jamal.

Jamal woke up 60 minutes before Li.

Li woke up 5 minutes after David.

David woke up 5 minutes before Enid.

Enid woke up 25 minutes before Fred.

Fred woke up 1 hour before Jorge.

Jorge woke up at 8:30 A.M.

1. Write the time each person went to sleep.

Name	Time
Annie	
Jamal	
Li	
David	
Enid	
Fred	
Jorge	

2. Write the time each person woke up.

Name	Time
Annie	
Jamal	
Li	
David	
Enid	
Fred	
Jorge	

Use the completed tables to answer each question.

3. Who woke up at 8:00 A.M.? ______________

4. Who woke up at 7:00 A.M.? ______________

5. Who slept the shortest time? ______________

6. Who slept the longest time? ______________

7. Who went to bed at 10:15 P.M.? ______________

8. Who slept 8 hours 55 minutes? ______________

Name ______________________________ Date ______________________________

LOGIC

Eliminating Wrong Choices

One of the keys to logic is the elimination of wrong choices. Sometimes, knowing what something *isn't* can help you find out what it *is.*

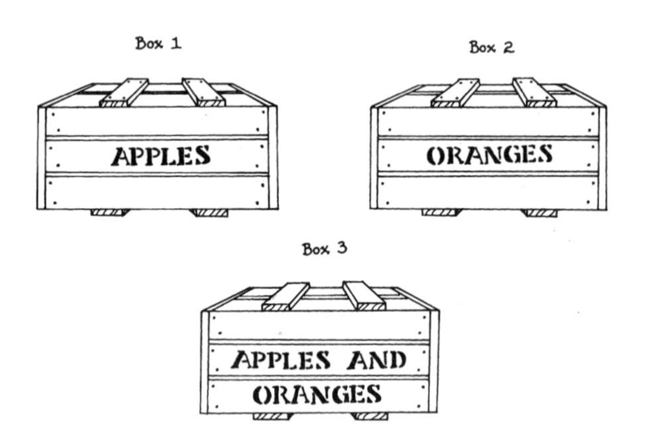

These three boxes were incorrectly labeled. None of them contains what is printed on the outside. However, it is possible to discover what is in each box by taking only one piece of fruit from only one of the boxes.

1. You know what is *not* in each box. Under each box write what *can be* in each box.

2. Suppose you reached into the box labeled *APPLES AND ORANGES* and pulled out an apple. What would that tell you? ______________________________

3. How would you know this? ______________________________

4. If the *APPLES AND ORANGES* box contains only apples, what must be in the *ORANGES* box? Explain. ______________________________

____

5. Then what must be in the *APPLES* box? why?

____

6. What would be in the other boxes if you found an orange in the *APPLES AND ORANGES* box?

Box 1: ______________________________ Box 2: ______________________________

So you see, sometimes what it isn't tells you what it is!

Name ______________________ Date ______________________

PATTERNS

Unit 4: Sequence

A **sequence** is a group of numbers that is formed by using a rule. In certain sequences, a constant number is added or subtracted from the number that goes before it. In other sequences, you divide or multiply the numbers by another number. In the sequence

1, 4, 7, 10, 13, 16,...

each number is 3 greater than the number before it.

$\underline{1} + 3 = \underline{4}$ $\underline{4} + 3 = \underline{7}$ $\underline{7} + 3 = \underline{10}$,   and so on.

Write the missing numbers in each sequence. Write fractions in simplest form.

1. 2, 5, ______, 11, ______, . . .

2. $7\frac{1}{2}$, 6, ______, 3, ______, . . .

3. 1.5, 2.25, ______. 3.75, ______, . . .

4. $15\frac{5}{8}$, ______, $22\frac{3}{8}$, $25\frac{3}{4}$, ______, . . .

In Examples 1 through 4, you added or subtracted a certain number to find the next number in the sequence. These sequences are called **arithmetic** sequences. Another type of sequence is a **geometric** sequence, in which you multiply or divide to find the next number. In the sequence

1.5, 3, 6, 12, 24,...

you multiply each number by 2 to get the next number in the sequence.

Write the missing number in each sequence. Write fractions in simplest form.

5. 2, 6, 18, ______, ______, ______, . . .
6. 43.9, 175.6, ______, ______, 11,238.4, . . .
7. $4\frac{1}{2}$, $6\frac{3}{4}$, $10\frac{1}{8}$, __________, __________ . . .

In some sequences, you might need more than one operation to find the next number in the sequence.

8. Write the next three numbers of this sequence. To find the next number, add 3 and then multiply by 0.5.

4.5, ______ , ______, ______, ...

Name ______________________ Date ______________________

PATTERNS

Divisible by 3

Is 3,533,772 divisible by 3? You can find out without dividing. Just do some simple addition.

First, add all of the single digits in the number.

$$3 + 5 + 3 + 3 + 7 + 7 + 2 = 30$$

If the sum of the digits in the original number is a 2-digit number, add the digits in the sum.

$$3 + 0 = 3$$

Is the final sum 3, 6, or 9? If it is, then the original number can be divided by 3.

$$3 \overline{)3{,}533{,}772} = 1{,}177{,}924$$

Complete the table. Write the sum of the digits in the number. Write *yes* if the number is evenly divisible by 3, and *no* if it is not. Find the quotient if the number is divisible by 3.

Number	Sum of digits	Divisible by 3?	Quotient
1. 1,986			
2. 33,333,304			
3. 45,188,010			
4. 19,561,522			
5. 15,303,401			
6. 16,301			
7. 403,482			
8. 602,730,770			
9. 102,560			
10. 3,973,485,621			

Name ______________________ Date ______________________

PATTERNS

Mathematical Sequence

A **mathematical sequence** is a group of numbers arranged to follow one another in a logical pattern. For example, the order of numbers 2, 4, 6, 8, 10, 12,... is a sequence of numbers divisible by 2.

You can make sequences with a row of numbers and their divisors to learn more about division with decimals.

1. Make a sequence with the following numbers. Arrange them in order from the least to the greatest. Then divide each number by 3. Place each quotient below its dividend.

2,421, 2,418, 2,427, 2,409, 2,415, 2,403, 2,406, 2,412, 2,424

______, ______, ______, ______, ______, ______, ______, ______, ______

______, ______, ______, ______, ______, ______, ______, ______, ______

1. Is there a logical sequence in the order of the quotients?

____

2. What is the relationship of the sequence of numbers in the middle row of Problem 1?

____

3. What is the relationship of the sequence of numbers in the bottom row of Problem 1?

____

4. Complete the tables by filling in the missing quotients. Can you predict what the answers will be before doing the arithmetic?

Dividends (columns); Divisors (rows)

÷	31,224	31,212	31,200	31,188
2				
3				
4				
6				

Dividends (columns); Divisors (rows)

÷	32,412	32,400	32,388	32,376
2				
3				
4				
6				

Name ______________________ Date ______________________

Modular System

One kind of number system is called a **modular system**. A modular system is best understood as the kind of arithmetical operations that we do "around a clock." Look at the face of a 5-hour clock. When you add 2 + 4, you move the hand on the clock 6 hours forward in a clockwise direction. In clock mathematics, or modular 5, 2 + 4 = 1.

Because multiplication can be thought of as repeated addition, we can say 4 x 4 = 4 + 4 + 4 + 4 = 16 = 1 (mod-5).

Complete the multiplication table for the modular-5, or mod-5, system.

×	0	1	2	3	4
0	0	0			
1	0	1			
2	0	2			
3	0	3			
4	0	4			

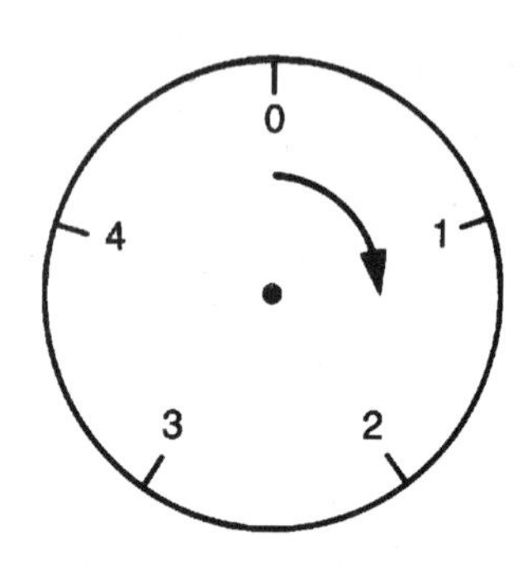

Complete the multiplication table for the modular-10 system.

×	0	1	2	3	4	5	6	7	8	9
0										
1										
2										
3										
4										
5										
6										
7										
8										
9										

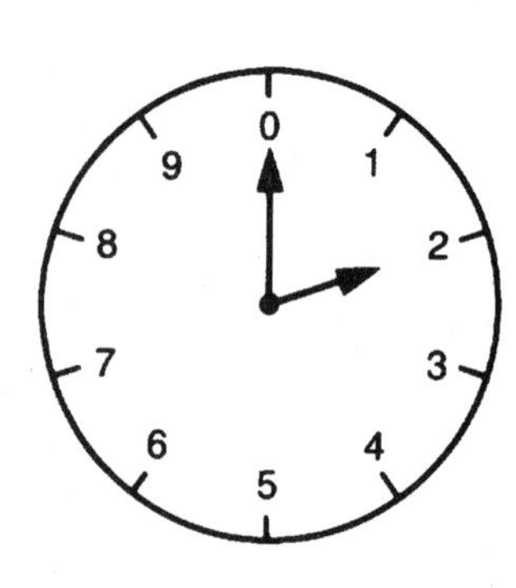

Name ______________________________ Date ______________________________

PATTERNS

Progression

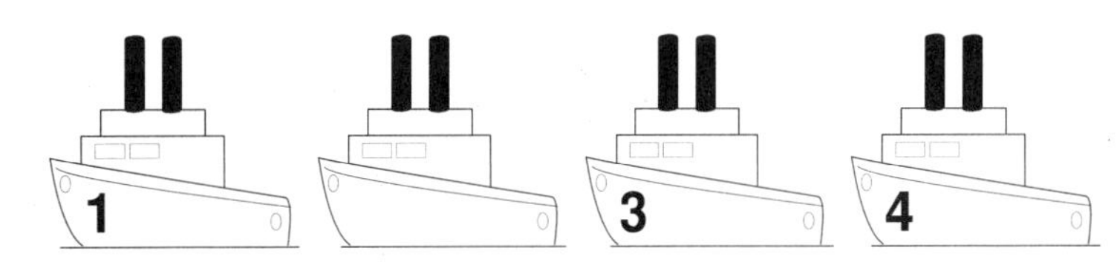

A **progression** is a sequence of related numbers. Each number in the progression is called a *term.* In an arithmetic progression, the new terms are found by adding the previous term. To find that fixed number, you find the difference between terms.

$\frac{1}{2}$, 1, $1\frac{1}{2}$, 2, $2\frac{1}{2}$, 3

1. What is the difference between each term in the progression?

____

2. What are the seventh, eighth, and ninth terms?

____

You can use equivalent fractions to help find the missing terms in a progression of fractions. Look at this sequence of fractions:

$\frac{1}{4}$, $\frac{9}{12}$, $\frac{10}{8}$

3. Reduce 1/4, 9/12, 10/8 to their lowest equivalent fractions.

____

4. Complete the progression.

$\frac{1}{4}$, $\frac{3}{4}$, $\frac{10}{8}$, $\frac{14}{\_\_}$, $\frac{36}{\_\_}$, $\frac{\_\_}{16}$, $\frac{104}{\_\_}$, $\frac{\_\_}{32}$

5. Complete the progression.

$\frac{1}{3}$, $\frac{2}{3}$, $\frac{6}{6}$, $\frac{\_\_}{6}$, $\frac{20}{12}$, $\frac{\_\_}{12}$, $\frac{\_\_}{24}$, $\frac{\_\_}{24}$

6. How did you find the missing terms in the last progression?

____

7. Complete the progression.

$\frac{1}{3}$, $\frac{2}{3}$, $\frac{6}{6}$, $\frac{12}{\_\_}$, $\frac{\_\_}{9}$, $\frac{18}{\_\_}$, $\frac{\_\_}{3}$, $\frac{\_\_}{3}$, 3

8. How did you find the missing terms in the last progression?

____

Name ______________________ Date ______________________

PATTERNS

Multiplication Patterns

Multiplication can produce some surprising patterns. Here are a few.

Multiply to find the number pattern.

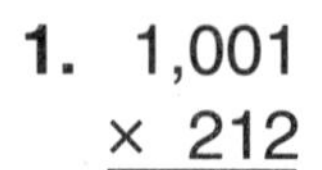

1. 1,001 × 212

2. 1,001 × 805

3. 1,001 × 479

4. How would you describe this pattern?

____

Multiply to find another number pattern.

5. 999 × 3

6. 999 × 9

7. 999 × 5

8. Describe this pattern.

____

____

Multiply to find a third number pattern.

9. 15,873 × 7

10. 15,873 × 14

11. 15,873 × 28

12. Describe this pattern.

____

____

Use the patterns that you have learned to solve each problem mentally.

13. 15,873 × 21

14. 1,001 × 567

15. 15,873 × 63

16. 999 × 8

Name ____________________ Date ____________________

Facts About Numbers and Their Factors

Here are some interesting facts about numbers and their factors.

Two numbers are said to be **friendly** or **amicable** if the sum of the proper factors of one number equals the other number, and vice versa. The **proper factors** of a number are all of the factors except the number itself. For example, an amicable pair of numbers is the pair 1,184 and 1,210.

Factors of 1,184: 1, 2, 4, 8, 16, 32, 37, 74, 148, 296, 592
Sum: 1 + 2 + 4 + 8 + 16 + 32 + 37 + 74 + 148 + 296 + 592 = 1,210

Factors of 1,210: 1, 2, 5, 10, 11, 22, 55, 110, 121, 242, 605
Sum: 1 + 2 + 5 + 10 + 11 + 22 + 55 + 110 + 121 + 242 + 605 = 1,184

The sum of the factors of 1,184 = 1,210.
The sum of the factors of 1,210 = 1,184.

1. Amicable pairs of numbers are very rare. The ancient Greeks were interested in amicable pairs of numbers. They were able to find only one pair. One number in this pair is 220. What is the other number?

____

The ancient Greeks also thought that certain numbers were magical. These were numbers that equal the sum of their proper factors. We call these numbers **perfect numbers**. For example, 6 is a perfect number.

Factors of 6: 1, 2, 3
Sum: 1 + 2 + 3 = 6

2. Find the only perfect number between 20 and 30. ________

Abundant numbers are numbers whose proper factors add up to sums greater than the number itself. For example, 12 is an abundant number.

Factors of 12: 1, 2, 3, 4, 6
Sum: 1 + 2 + 3 + 4 + 6 = 16

3. Find the other two abundant numbers between 10 and 21.

____

Name ______________________ Date ______________________

PATTERNS

Tangram

A **tangram** is a Chinese puzzle made from a square, a parallelogram, and triangles. They can be put together to make a larger square or different shapes.

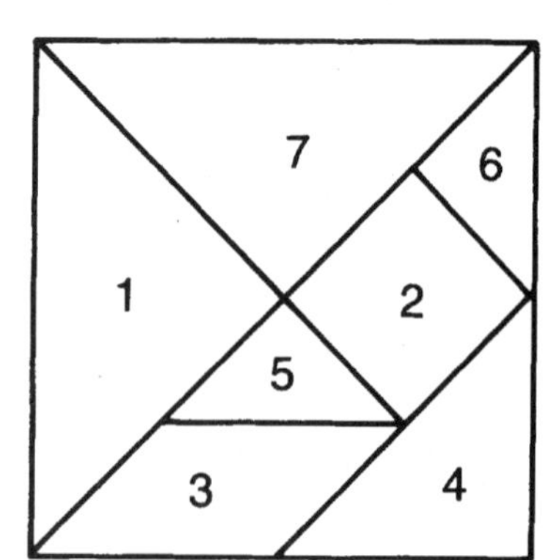

a. Use tracing paper to trace the tangram. Number each part as shown.

b. Cut out each part so that you can move the parts to form new shapes.

c. In the table, draw a sketch of how you put the pieces together. Copy the numbers. All the shapes are possible but one.

	Shapes to make	(square)	(triangle)	(parallelogram)
Number of polygons to use	2	5 6	5 6	5 6
	3			
	4			
	5			
	6			
	7			

Name ______________________ Date ______________________

Congruent Figures

Two figures of the same size and shape are **congruent figures**. Some figures can be divided into congruent halves. You can place one half on top of the other so that both have the same outline.

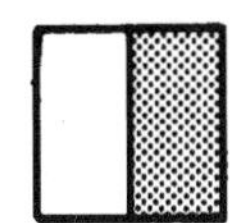      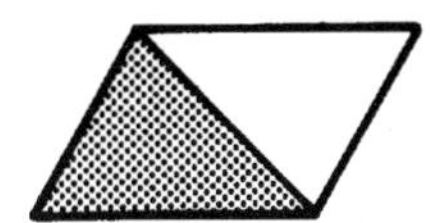

Some figures have congruent halves that are harder to find. Their two halves are not back to back. Look at the three figures with congruent halves.

  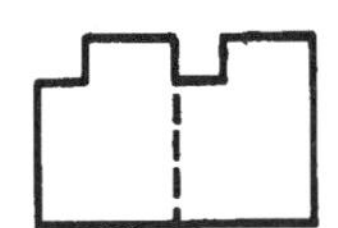

Find the congruent halves of each figure. Use a line to separate each of the congruent halves.

1. 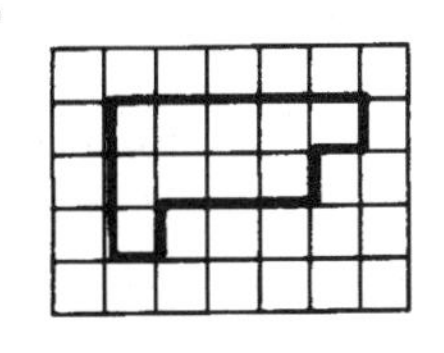

2. 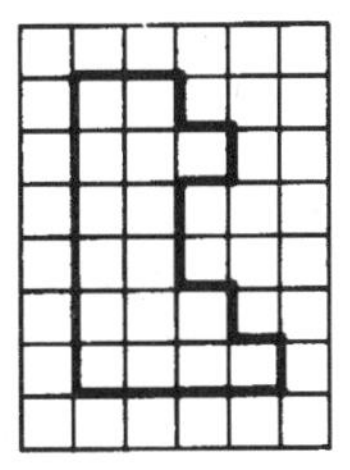

3. 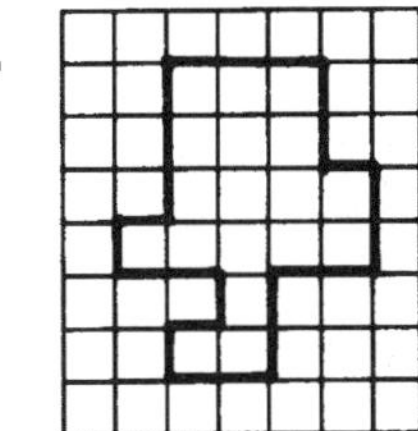

4. 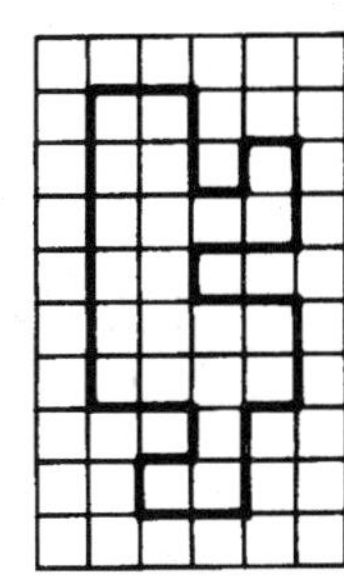

5. 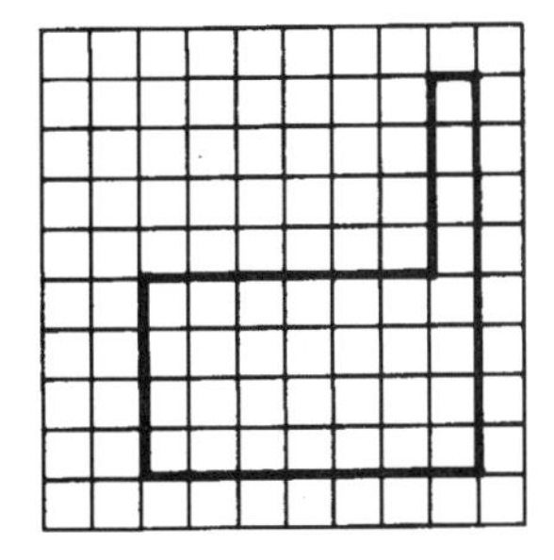

6. 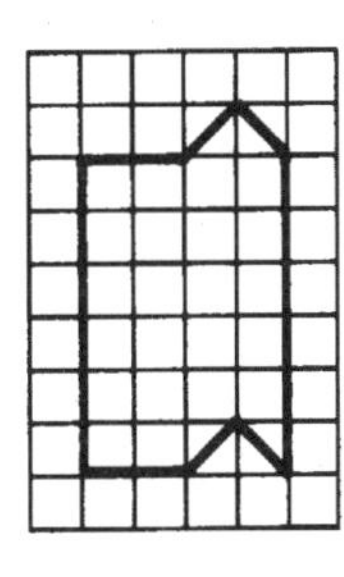

7. 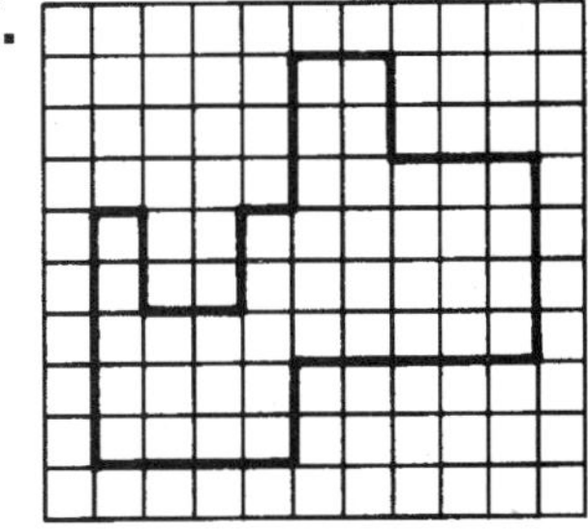

8. 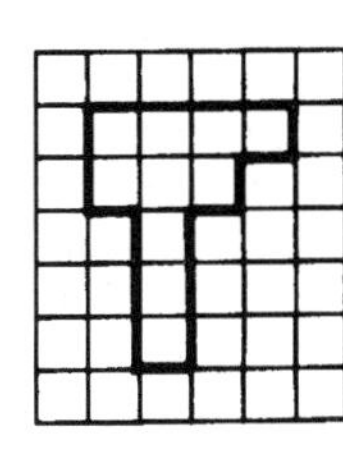

9. 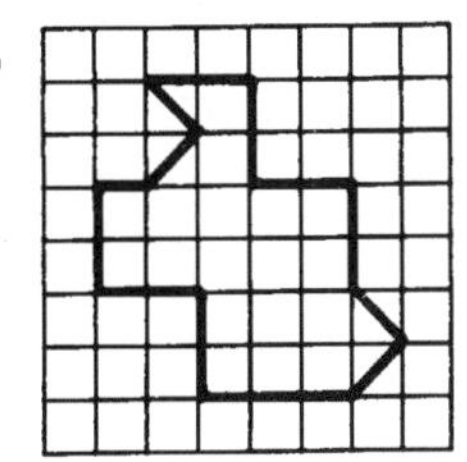

10. 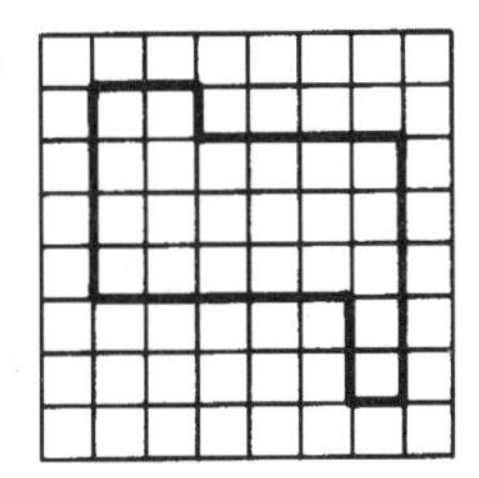

11. 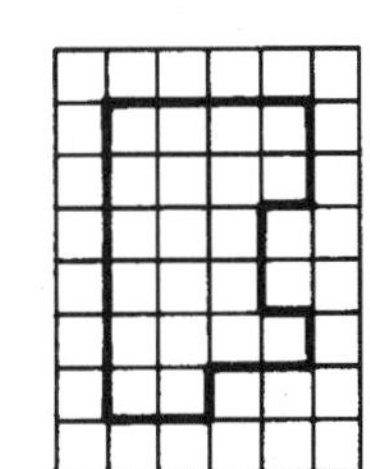

12. 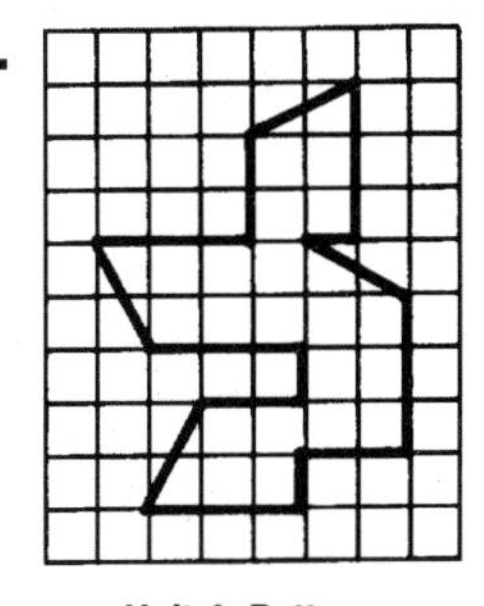

Name ______________________ Date ______________________

PATTERNS

Different Angles

Study the first two cubes. They show how the same cube would look if it were seen from different angles. In each problem, study the left cube carefully. Then fill in the missing letters the way they would be shown if the cube were seen from a different angle. The same three sides are visible on cubes.

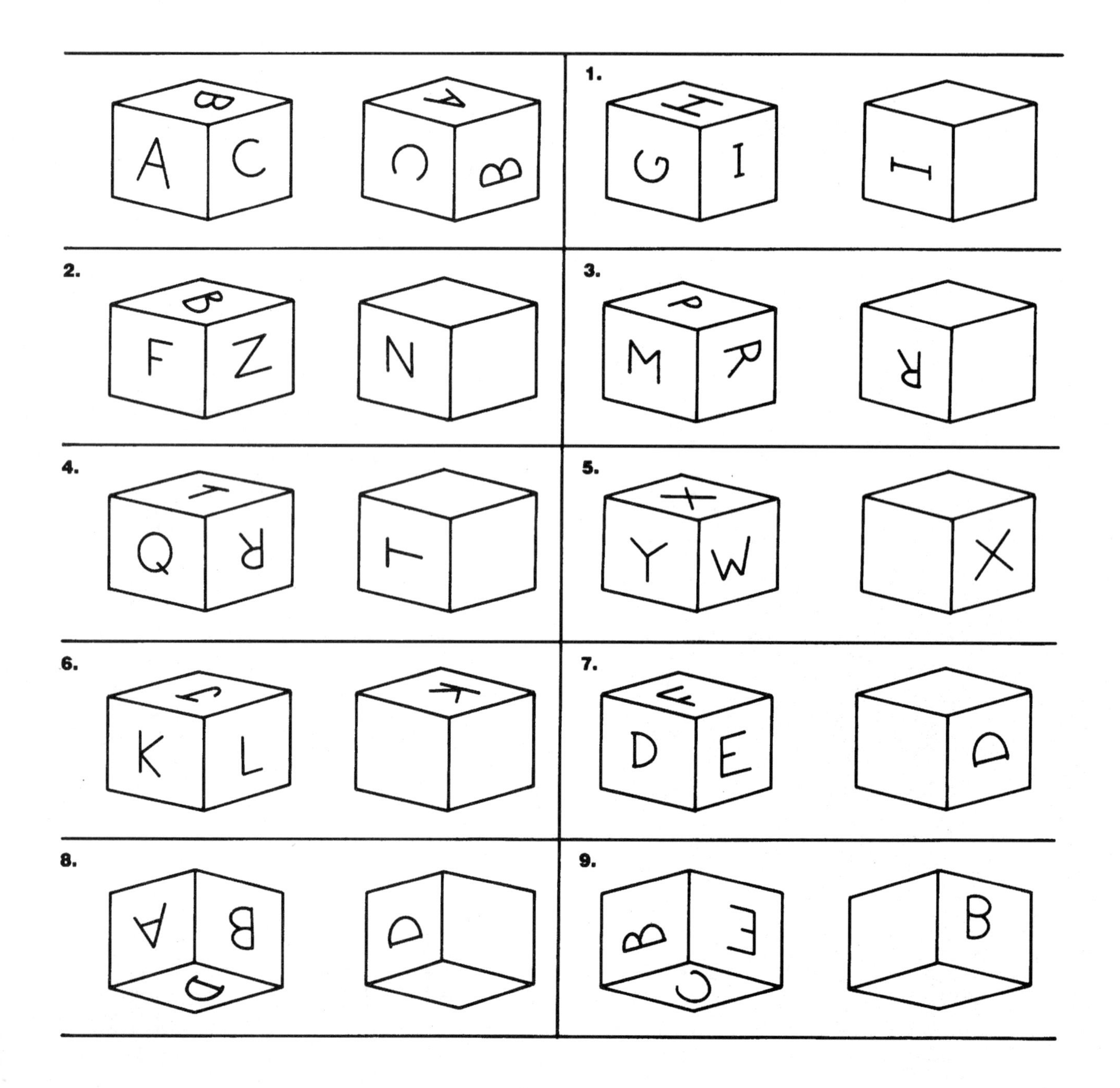

Name ______________________ Date ______________________

ALGEBRA

Unit 5: Related Facts

Read each problem. Each symbol stands for a number. Write *add* or *subtract* to tell how you would solve each problem.

1. Sasha has □ flyers. She hands out △ flyers. How many flyers does she have left to give out?

____

2. Jill counts ▱ voters in the first hour. She counts ◺ during the next hour. How many does she count in all?

____

3. Lamont spent ⬡ on this election. He will spend ✳ the next time he runs. How much more will he spend?

____

4. There are ⧋ people in the neighborhood. Out of those people, ⛉ did not vote. How many people voted?

____

5. Lamont Brown received ⌂ votes, and Sharon Vargas received ◸ votes. How many votes did Sharon lose by?

____

6. Jim received ○ buttons to hand out. He has ⊖ buttons left. How many buttons did he give away?

____

Fill in each box with one of the three numbers. Then write + or - in the circles to write a true number sentence. The first one is done for you.

7. 18 14 1

18 (+) 1 (−) 14 = 5

8. 12 5 3

□ ○ □ ○ □ = 14

9. □ ○ □ ○ □ = 8

10. □ ○ □ ○ □ = 7

11. □ ○ □ ○ □ = 4

12. □ ○ □ ○ □ = 3

13. □ ○ □ ○ □ = 12

14. □ ○ □ ○ □ = 6

Name ______________________ Date ______________________

ALGEBRA

Adding Decimals

A firm uses secret formulas to make six different perfumes. When someone spills grape juice over the formulas, some of the numbers are blotted out. Use arithmetic to complete each formula.

1.
```
   1 . □ 0 6
+  0 . 7 □ □
   □ . 7 4 8
```

2.
```
   □ . 8 9
+  2 . □ □
   9 . 3 5
```

3.
```
   2 4 . 0 □
+  1 □ . □ □
   □ 0 . 2 0
```

4.
```
   3 . 4 6 □
+  □ . 0 □ 8
   3 . □ 1 0
```

5.
```
    9 . 3 □ □
+   □ . □ 4 7
   1 0 . 3 0 5
```

6.
```
   □ 2 . 6 4
+  1 □ . 5 □
   6 3 . □ 1
```

In another mishap, more formulas were partly blotted out. Each of these formulas contained every number from 1 to 9. Fill in all the missing numbers. For each formula, list the numbers 1 through 9 on a piece of paper. Then cross out each that has already been used.

7.
```
   □ □ . 9
+  □ 3 . □
   5 □ . □
```

8.
```
   □ . 1 □
+  □ . □ 6
   9 . □ □
```

9.
```
   □ . 5 □
+  6 . □ □
   □ . □ 3
```

10.
```
   □ 8 . □
+  3 □ . □
   □ □ . 5
```

11.
```
   7 . □ □
+  1 . □ 2
   □ . 3 □
```

12.
```
   2 □ . □
+  □ 5 . □
   □ □ . 3
```

13. You may have discovered that in Exercises 7 to 12 there is more than one correct answer for each formula. Why?

____

____

____

____

Name ______________________ Date ______________________

ALGEBRA

Properties of Multiplication

You know there are five basic properties of multiplication.

a. **The Identity Property**—Any factor multiplied by 1 equals that number. $23 \times 1 = 23$

b. **The Zero Property**—Any factor multiplied by zero equals 0. $3 \times 0 = 0$

c. **The Distributive Property**—To find the product of a number times the sum of addends, you can multiply each addend by the number, then add the products. $4 \times (7 + 1) = (4 \times 7) + (4 \times 1)$

d. **The Associative Property**—The way factors are grouped does not affect the product. $5 \times (3 \times 2) = (5 \times 3) \times 2$

e. **The Commutative Property**—The product is not affected by the order of the factors. $18 \times 2 = 2 \times 18$

Find the number value of *n* in each multiplication problem, and fill in on the lines to the right the multiplication property it demonstrates.

1. $9 \times (7 \times n) = (9 \times 7) \times 5$ $n =$ ___ _ _ _ _ _ _ _ _ _ _ _ _ (★ under 6th blank, ό under 7th)
2. $9 \times (4 + 6) = (9 \times n) + (9 \times 6)$ $n =$ ___ _ _ _ _ _ _ _ _ _ _ _ _ (☺ under 9th)
3. $4 \times n = 4$ $n =$ ___ _ _ _ _ _ _ _ _ _ _ _ _ (✳ under 2nd, ナu under 4th)
4. $7 \times 4 = n \times 7$ $n =$ ___ _ _ _ _ _ _ _ _ _ _ _ _ (日 under 2nd, ൭ under 9th)
5. $n \times (6 + 5) = (4 \times 6) + (4 \times 5)$ $n =$ ___ _ _ _ _ _ _ _ _ _ _ _ _ (☼ under 1st)
6. $1 \times n = 1$ $n =$ ___ _ _ _ _ _ _ _ _ _ _ _ _ (ん under 2nd, ☾ under 5th, ⋐ under 7th)
7. $n \times 3 = 3 \times 8$ $n =$ ___ _ _ _ _ _ _ _ _ _ _ _ _ (△ under 7th)
8. $5 \times (3 + n) = (5 \times 3) + (5 \times 2)$ $n =$ ___ _ _ _ _ _ _ _ _ _ _ _ _ (上 under 1st)
9. $6 \times n = 9 \times 6$ $n =$ ___ _ _ _ _ _ _ _ _ _ _ _ _

To solve the riddle, fill in each blank with the letter identified by the symbols in the answers from the problems above.

Q.: Why did Sherlock Holmes throw out his calculator?

A.: Because _ _ _ _ _ _ _ _ _ _ _ _ up

★ ☺ ☼ ൭ 上 ナu 日 ☾ ό ✳ ん

Name ______________________________ Date ______________________________

ALGEBRA

Determine the Operation

Each of these equations has a different symbol to show which math operation you should use. It's your job to find what the symbol means.

Example:

2 ● 16 ● 5 = 160

Is this operation addition? Try it and see.

2 + 16 + 5 = 23

Is the operation multiplication? Try it and see.

2 × 16 × 5 = 160

Multiplication works for this problem. ● means "multiply."

Find each correct operation.

1. 12 □ 85 □ 97 = 194

□ means you ________________.

2. 980 ~ 4 ~ 5 ~ 7 = 7

~ means you ________________.

3. 874 △ 533 △ 35 △ 69 = 237

△ means you ________________.

4. 65 ○ 36 ○ 2 = 4,680

○ means you ________________.

5. 2 ◇ 85 ◇ 58 = 9,860

◇ means you ________________.

6. 820 ▱ 205 ▱ 2 = 2

▱ means you ________________.

Solve each problem.

7. 11 δ 76 δ 200 = 287

3 δ 97 δ 376 = _____

8. 12,600 β 63 β 5 β 8 = 5

77,900 β 82 β 5 β 2 = _____

9. 1,243 ▲ 543 ▲ 68 ▲ 100 = 532

9,543 ▲ 2,551 ▲ 3,800 ▲ 999 = ___

10. 190 ▼ 190 ▼ 654 = 23,609,400

2 ▼ 44 ▼ 666 = _____

11. 32 ☆ 56 ☆ 101 = 180,992

673 ☆ 5 ☆ 5 ☆ 31 = _____

12. 12,285 ∂ 65 ∂ 3 ∂ 9 = 7

47,628 ∂ 98 ∂ 6 ∂ 9 = _____

Name ______________________ Date ______________________

ALGEBRA

Proportion

A **recipe** shows how many people it is meant to serve. When it serves more people, then the amount of the ingredients must be changed in **proportion** to the number of people it will serve.

Change the amounts of the ingredients of each recipe.

Cheese Bake	*Cheese Bake*
(Serves 3)	(Serves 8)
1 c flour	_____ c flour
$1\frac{1}{2}$ tsp baking powder	_____ tsp baking powder
$\frac{1}{2}$ tsp salt	_____ tsp salt
2 tsp butter	_____ tsp butter
$\frac{1}{2}$ c grated cheddar cheese	_____ c grated cheddar cheese
$\frac{1}{3}$ c cold water	_____ c cold water

Potato Delight	*Potato Delight*
(Serves 4)	(Serves 14)
2 c bread crumbs	_____ c bread crumbs
4 eggs	_____ eggs
14 oz evaporated milk	_____ oz evaporated milk
1 tsp onion powder	_____ tsp onion powder
3 tsp salt	_____ tsp salt
$1\frac{1}{2}$ tbsp Worcestershire sauce	_____ tbsp Worcestershire sauce
4 potatoes	_____ potatoes
8 oz frozen vegetables	_____ oz frozen vegetables
5 raw carrots	_____ raw carrots
7 oz gravy	_____ oz gravy

Name ______________________ Date ______________________

ALGEBRA

Index Numbers

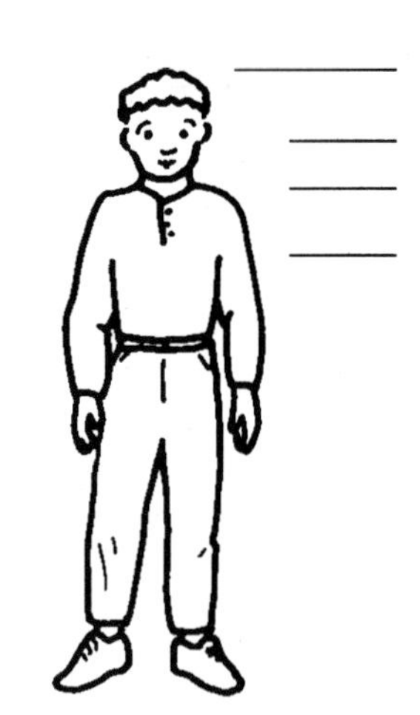

Index numbers are a series of numbers that measure amounts. Index numbers usually run from 0 to 100. An amount is chosen as the standard and is given the value of 100. Mr. Cohen, a science teacher, uses index numbers to compare his students' growth. This year, Mr. Cohen chose Ray's height as the standard. Ray was 50 inches tall at the beginning of the school year. Mr. Cohen gave this height a value of 100. He gave heights less than 50 inches values of less than 100. Heights greater than 50 inches were given values of more than 100. For example, Petra is 40 inches tall. To find the index number for 40 inches, you use ratio and proportion. The ratio of Petra's height to Ray's height is in proportion to the ratio of her index number and 100.

$$\frac{40}{50} = \frac{x}{100} \qquad \begin{aligned} 50x &= 40 \times 100 \\ 50x &= 4{,}000 \\ x &= 80 \end{aligned}$$

The index number for Ray's height is 100. The index number for Petra's height is 80. Complete the table.

Student	Height	Index Number
Ray	50 inches	100
Petra	40 inches	80
Les	inches	120
Denise	45 inches	
Freida	52 inches	
Luis	48 inches	
Josh	inches	124

1. Mr. Cohen is 72 inches tall. What is the index number for his height?

   ____

2. After two months, the new index number for Freida is 106. How much has she grown?

   ____

3. What percent of Ray's height is Luis's height? if Luis were 2 inches taller?

   ____

4. Ray grew 3 inches in one year. What percent did his height change?

   ____

Name ______________________ Date ______________________

MEASUREMENT

Unit 6: Light-Years

Space travelers from the planet Wong traveled across the universe to the brightest stars. Distances in outer space are so great that they are measured in light-years. A *light-year* is the distance light travels in one year—5.88 trillion miles.

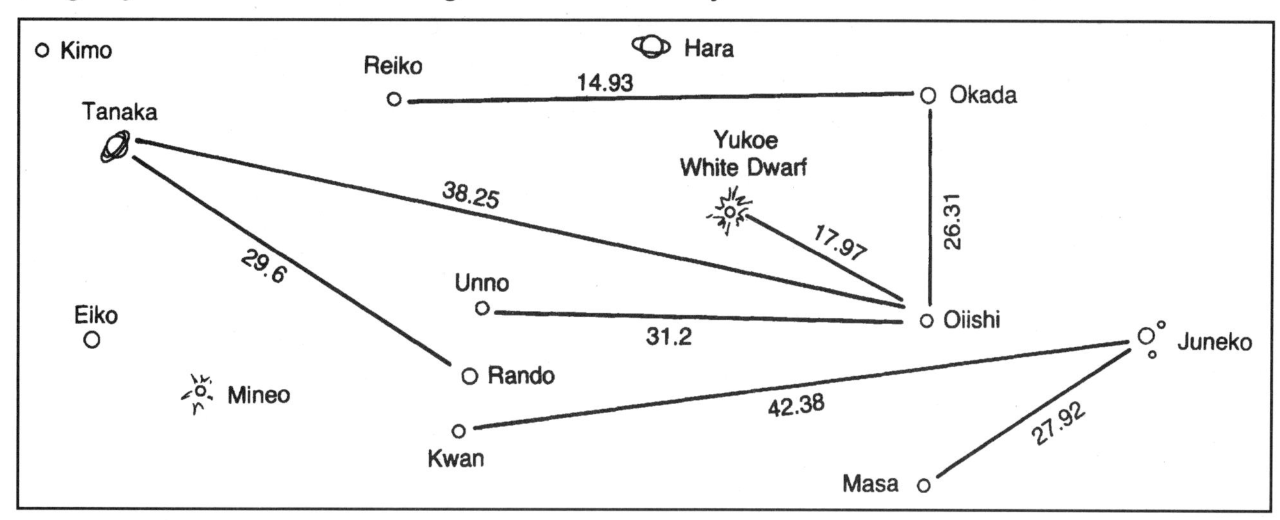

Write the answers.

1. It is 1 light-year from Hara to Okada. The space travelers have traveled 1 trillion miles of the way. How many miles remain before they reach Okada?

   ____

2. It is 38.2 light-years from Hara to Kimo, and 27.66 light-years from Hara to Tanaka. How much farther is it from Hara to Kimo than Hara to Tanaka?

   ____

3. How much greater is the distance between Unno and Oiishi than the distance between Okada and Reiko?

   ____

4. How much greater is the distance between Oiishi and Okada than between Oiishi and the Yukoe White Dwarf?

   ____

5. How much farther is it from Juneko to Kwan than from Juneko to Masa?

   ____

6. How much farther is it from Tanaka to Oiishi than from Tanaka to Rando?

   ____

7. You know the distance that light travels in 1 year. But what distance does it travel in a shorter period of time? How would you find the distance for 1 day? What is the distance?

   ____

Name ______________________ Date ______________________

MEASUREMENT

Type Size

Editor Needed

Editor Needed

Editor Needed

The type that you see in newspapers is printed in different sizes, or **points**. Large point sizes are used in headlines. Smaller point sizes are used in the stories. **Picas** are used to see how wide and how long a page of type is. A pica is equal to 12 points. Editors and printers have to know the number of characters of type per pica.

10 points = 2.5 characters per pica
A column that is 20 picas wide will have
$20 \times 2.5 = 50$ characters of 10-point type per line.

Editors also need to know how many lines can fit in a column. Type sizes and column lengths often vary; so, they use a formula for measuring. They multiply the number of picas in a column by 12. Then they divide by the point size of the type.

Column length: 60 picas
Type size: 10 points
60 picas $\times$ 12 = 720 points $\div$ 10 = 72 lines per column

Solve.

1. Your newspaper has columns that are 50 picas long. How many lines of copy can fit on a page of 10-point type? on a page of 12-point type?

   ____

2. You want to print an article set in 14-point type. Your page has columns that are 70 picas long. How many lines of 14-point copy will fit in a column?

   ____

3. You want to print a sports headline in 18-point type (1.4 characters per pica). The headline has 28 characters. The space for it is 15 picas wide. Will the headline fit?

   ____

4. You need to reduce the type size of an article. Each column is 65 picas. How many lines fit when the type is 15 points? How many lines fit after reducing the copy to 12-point type?

   ____

5. You are using 10-point type (2.5 characters per pica), and you need to write a story in a column that is 15 picas long and 20 picas wide. How many characters do you have to work with?

   ____

6. You are writing a sports story that will be printed in 18-point type (1.4 characters per pica). Your column is 30 picas long and 10 picas wide. How many characters do you have to work with?

   ____

Name ______________________ Date ______________________

MEASUREMENT

Lowest Common Multiple

Jenny is planning a birthday party for her younger brother. She needs to buy construction paper to make invitations and name tags for the guests. Each invitation will be 6 inches square. Each name tag will be 4 inches square. She wants to buy square sheets of paper in a single size only and cut those squares into invitations and name tags. She wants to be able to cut down the larger squares without wasting paper. What is the smallest square sheet of paper that Jenny can buy?

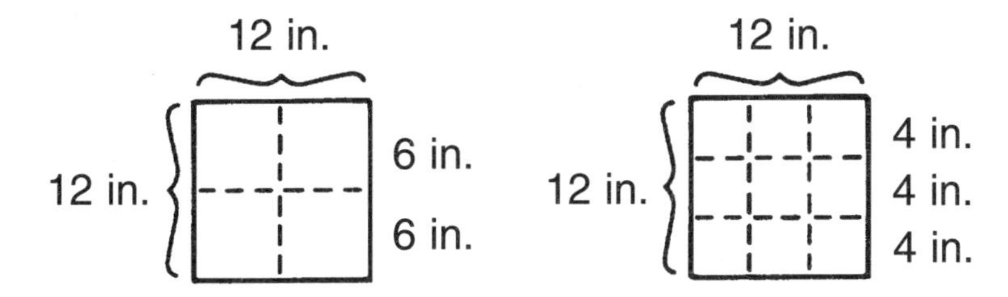

A sheet of construction paper 12 inches long and 12 inches wide can be cut into four 6-inch squares. The same sheet can be cut into nine 4-inch squares.

An easy way to find what size paper Jenny needs is to find the **lowest common multiple,** or **LCM**, of 4 and 6. List the first 6 multiples of 4 and the first 4 multiples of 6. 4 → 4, 8, 12, 16, 20, 24 6 → 6, 12, 18, 24

You can see that 12 and 24 are common multiples for 4 and 6. But 12 is the LCM for 6 and 4.

Find the smallest square sheet of paper that can be cut into pieces of paper for the sizes given.

1. 3 inches square or 5 inches square ______________
2. 27 inches square or 3 inches square ______________
3. 3 inches square or 8 inches square ______________
4. 6 inches square or 9 inches square ______________
5. 12 inches square or 20 inches square ______________
6. 36 inches square or 60 inches square ______________
7. 16 inches square or 32 inches square ______________

Name ______________________ Date ______________________

MEASUREMENT

Time Schedule

You plan to send a package from New York City to Bradenton, Florida. There is air freight service as far as Sarasota, Florida. From Sarasota to Bradenton, there is only bus service. All packages must arrive at the freight department of the New York City terminal 45 minutes before the flight. The ride from your home to the terminal takes 20 minutes.

AIRLINE & BUS FREIGHT SCHEDULE—DAILY

Air	Depart New York	6:30 A.M.	10:00 A.M.	1:30 P.M.	5:00 P.M.
Freight	Arrive Sarasota	9:30 A.M.	1:00 P.M.	4:30 P.M.	8:00 P.M.
Bus	Depart Sarasota	10:30 A.M.	2:00 P.M.	5:30 P.M.	9:00 P.M.
Freight	Arrive Bradenton	11:30 A.M.	3:00 P.M.	6:30 P.M.	10:00 P.M.

Write the answers.

1. To arrive at the airline terminal at 10:00 A.M., when should you leave your home? ______________________

2. Your package must go out on the 1:30 P.M. flight. When should you get to the terminal? ______________________

3. If you deliver the package to the airline freight department at 9:15 A.M., when will it reach Sarasota? When will it reach Bradenton? ______________________

4. You deliver your package to the airline freight department at 5:30 P.M. When will it arrive in Bradenton? ______________________

5. You want your package to arrive in Bradenton at 3:00 P.M. When must you leave your home? ______________________

6. The flight is 25 minutes late arriving in Sarasota. Will your package still arrive in Bradenton on time? ______________________

____

Name ______________________ Date ______________________

MEASUREMENT

Balance Scale

You can use a **balance scale** to measure the difference between the weights of objects. If the weight on the left is 1 gram and the weight on the right is 4 grams, then the scale will read 3 grams.

Use weights of 1 gram, 4 grams, 8 grams, and 16 grams to complete the table.

Left side	Scale reading	Right side
1 g	7 g	
8 g		16 g
4 g	12 g	
	4 g	4 g + 16 g
4 g + 4 g	0	
	3 g	1 g + 4 g
8 g + 16 g	19 g	
	27 g	1 g
	5 g	16 g + 1 g
16 g + 4 g	7 g	
16 g	13 g	
	11 g	
	13 g	8 g
	11 g	
28 g	1 g	

Name ______________________ Date ______________________

MEASUREMENT

Fraction of Total Hours

Five fifth-grade students have weekend paper routes in order to earn extra money. They made a table to show how many hours they spend doing different tasks.

Complete the table.

Students	Total hours during weekend	Fraction of total hours		
		Folding newspapers	Selling subscriptions	Delivering newspapers
Robert	$6\frac{2}{3}$	$\frac{1}{3}$		$\frac{1}{2}$
Ellen	$5\frac{1}{4}$	$\frac{1}{6}$	$\frac{1}{6}$	
Harold	$8\frac{1}{3}$	$\frac{1}{5}$		$\frac{2}{5}$
Lucia	9		$\frac{1}{6}$	$\frac{1}{2}$
Steve	$12\frac{1}{2}$	$\frac{3}{8}$	$\frac{1}{8}$	

Use the table to write the answers.

1. How many hours does each student spend delivering newspapers?

____

2. Who spends more time delivering newspapers, Ellen or Lucia?

____

3. Does Robert or Ellen spend more time folding newspapers?

____

4. If each student's hours are the same every weekend, how many hours does each student work in a month?

____

Name ______________________ Date ______________________

MEASUREMENT

Water Clock

A water clock was invented in the Middle Ages. Water would drip into a tank inside the clock, causing the water level to rise. A device floating on the water was connected to the hands of the clock. As the tank filled, the hands of the clock revolved. When 24 hours had passed, the tank would empty and the cycle would begin again.

Suppose you have a water clock that holds 20 gallons.

1. How many hours have passed when the tank contains 10 gallons of water? How full is the tank?

____

____

2. How many hours have passed when the tank contains 5 gallons of water?

____

3. How many hours have passed when the tank is $\frac{3}{4}$ full?

____

4. If the tank is empty at midnight, how many gallons are there in the tank 21 hours later? What time of day is it?

____

5. How many hours has the clock run after 45 gallons of water have been used? How many days is this?

____

Name ______________________ Date ______________________

MEASUREMENT

Map Scale

The **scale** on a map is the relationship between distance on the map and actual distance. It is usually written as a **ratio**. For example, a scale of 1 inch to 3 miles would mean that 1 inch on the map is equal to 3 miles of actual distance. The ratio in this scale would be written 1:3, or $\frac{1}{3}$. On this map, the scale is written in the lower right corner.

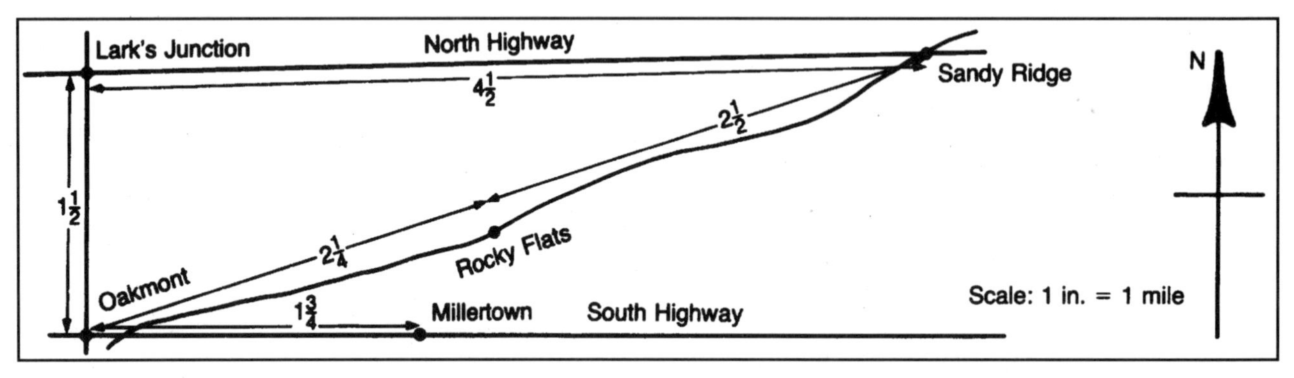

Use the map to answer the questions.

1. At Oakmont, North Highway and South Highway are $1\frac{1}{2}$ inches apart. How far apart are they actually?

   ____

2. Lark's Junction and Sandy Ridge are $4\frac{1}{2}$ inches apart. How far apart are they actually?

   ____

3. If you traveled from Lark's Junction to Oakmont, then from Oakmont to Sandy Ridge, how far would you go?

   ____

4. Which towns would you pass through if you took the shortest route from Sandy Ridge to Millertown? How far would you go?

   ____

Maps have many different scales. 1 inch on the map might equal several miles of actual distance. Suppose that the map above had a scale of 1 inch = 50 miles.

5. How far apart would North Highway and South Highway actually be?

   ____

6. How far apart would Sandy Ridge and Oakmont actually be?

   ____

7. How far apart would Oakmont and Millertown actually be?

   ____

8. What would be the actual distance from Lark's Junction to Sandy Ridge?

   ____

Name ______________________ Date ______________________

MEASUREMENT

Celsius Scale

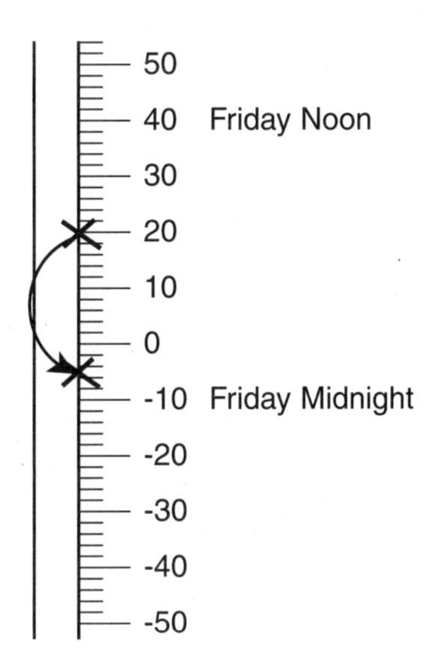

At zero degrees on the Celsius scale, water turns into ice. Any temperature warmer than zero is shown by numbers like 5° or 20°. Any temperature colder than zero is shown by numbers like -5° or -20°. Zero separates numbers like 1, 2, 3, 4 from numbers like -1, -2, -3, -4.

Last winter the Carters rented a ski cabin in the mountains. The thermometer showed 20° when the Carters arrived on Friday at noon. By midnight, the thermometer showed -5°. The temperature had fallen by 25°.

Show how the temperature changed on each of the next four days. Mark an X by both temperatures on the thermometer. Then write how many degrees the temperature changed from noon to midnight.

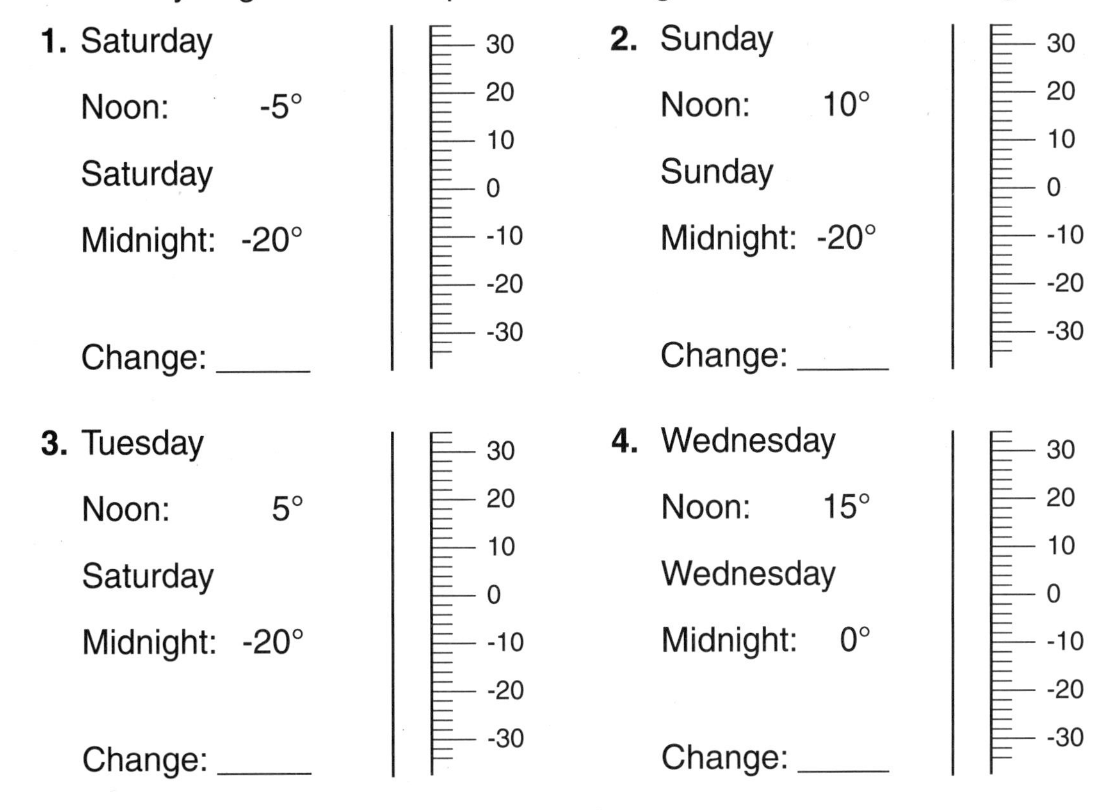

1. Saturday

Noon: -5°

Saturday

Midnight: -20°

Change: _____

2. Sunday

Noon: 10°

Sunday

Midnight: -20°

Change: _____

3. Tuesday

Noon: 5°

Saturday

Midnight: -20°

Change: _____

4. Wednesday

Noon: 15°

Wednesday

Midnight: 0°

Change: _____

During the next week, the Carters noted the daytime temperature and how it had changed by nighttime. What were the exact temperatures each night?

5. Friday Noon: -2°
It became colder by 6°.

Friday Midnight: ______

6. Sunday Noon: 14°
It became colder by 14°.

Sunday Midnight: ______

Name ____________________ Date ____________________

MEASUREMENT

Ratio

Ralph loves vegetable soup. He knows how much of each ingredient he needs to serve 2 people. The card shows his recipe.

Ralph's Vegetable Soup	
2 oz mushrooms	$\frac{1}{2}$ lb broccoli
1 carrot, sliced	1 stalk celery, chopped
$\frac{3}{4}$ tsp salt	
$\frac{1}{2}$ onion	3 slices zucchini

Look at the recipe card and solve.

1. How many onions will Ralph need to make soup for 6 guests?

   ____

2. If Ralph makes soup for 7, how much broccoli does he need?

   ____

3. Ralph bought 6 carrots. If he makes his soup for 6 people, how many carrots will he have left?

   ____

4. Ralph uses 1 quart of water when he cooks soup for 2. How much water will he use if he makes soup for 4 people?

   ____

5. Ralph doubled the amount of celery he usually puts in soup for 2 people. How much celery will he use if he cooks for 8 people?

   ____

6. How much salt will Ralph use if he makes vegetable soup for 4 people?

   ____

7. Complete the ratio table.

	Mushrooms	Carrots	Salt	Onions	Broccoli	Celery	Zucchini
2 people	2 oz	1	$\frac{3}{4}$ tsp	$\frac{1}{2}$	$\frac{1}{2}$ lb	1 stalk	3 slices
6 people							

Name ______________________ Date ______________________

MEASUREMENT

Currency

Different countries use different kinds of money. The money that a country uses is called its *currency.* To buy things in other countries, you first have to exchange your currency for the currency of that country. The table shows how much foreign currency you can receive in exchange for one United States dollar.

Country	Name of currency	Amount received for $1 U.S.
Saudi Arabia	riyal	3.65
Spain	peseta	164.70
Egypt	pound	1.30

Country	Name of currency	Amount received for $1 U.S.
Greece	drachma	131.85
Mexico	peso	330.00
Norway	krone	8.24

You can find how much foreign currency you can exchange for U.S. dollars by multiplying. From the table, you see that $1 = 3.65 riyal. To find how many riyal you receive for $25: $25 \times 3.65 = 91.25$

You can find the amount of U.S. dollars you will receive in exchange for foreign currency by dividing. To find how many dollars you will receive for 57.68 krone: $57.68 \div 8.24 = 7$

You must divide the larger amount by the smaller amount equal to $1 U.S.

1. Björn thinks that 182.5 riyal is worth more in dollars than 7,411.5 pesetas. Is he right?

   ____

2. Lena exchanged 39.0 Egyptian pounds for Greek drachmas. How many drachmas did she receive?

   ____

3. Complete the chart to show how much money Rick spent on his trip around the world.

Country	Mexico	Saudi Arabia	Egypt	Greece	Spain	Norway
Foreign currency			67.6 pounds		625.86 pesetas	
U.S. dollars	$55	$70		$69		$63

Name ______________________ Date ______________________

MEASUREMENT

Double-Bar Graph

Double-bar graphs are used to compare two sets of information that cover the same period of time. When the student councils at Pine Ridge School and Marble Creek School had a 4-week newspaper drive, they used a double-bar graph to show the number of pounds they collected each week.

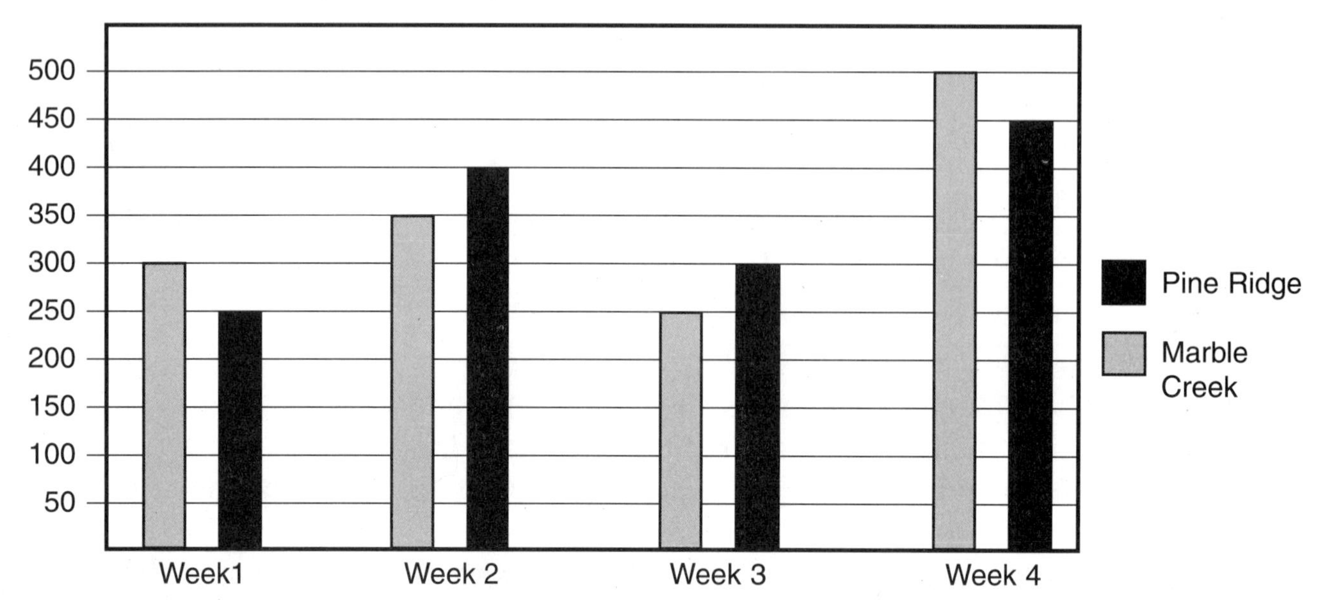

Solve the problems by using the graph.

1. How many pounds of newspaper were collected by both groups in Week 1?

____

2. Which group collected more pounds of newspaper in Week 2?

____

3. Which group collected more pounds of newspaper in Week 3?

____

4. Which group collected more pounds of newspaper in Week 4?

____

5. In which week did the groups collect the most newspaper altogether?

____

6. How many pounds were collected by each group that week?

____

Name ______________________ Date ______________________

MEASUREMENT

Pictographs

Look at the pictograph for the average daily hours of sunlight in Alaska. It shows the average hours of sunlight for days in spring, summer, fall, and winter.

Average Daily Hours of Sunlight in Alaska

Spring	
Summer	
Fall	
Winter	

= 2 hours
= $1\frac{1}{2}$ hours
= 1 hour
= $\frac{1}{2}$ hour

Use the pictograph to write the answer.

1. How many hours of sunlight are there on a day in spring? in winter?

____

2. How many hours of darkness are there on a fall day? on a winter day?

____

3. How many more hours of sunlight are there on a summer day than on a spring day?

____

4. How many more hours of darkness are there on a winter day than on a fall day?

____

Use the following inches of average rainfall to complete the pictograph.

July .... $8\frac{1}{4}$ August .... $6\frac{3}{4}$ September .... $7\frac{1}{4}$

October .... 4 November .... $1\frac{1}{2}$ December 2

January		July	
February		August	
March		September	
April		October	
May		November	
June		December	

= 1 inch = $\frac{3}{4}$ inch = $\frac{1}{2}$ inch = $\frac{1}{4}$ inch

Name ______________________ Date ______________________

MEASUREMENT

Broken-Line Graph

The Roxy movie theater has 250 seats. Attendance at the theater varies from night to night. Sometimes it is full. Sometimes only 100 seats are sold. In one 2-week period, the manager made a **broken-line graph** of attendance.

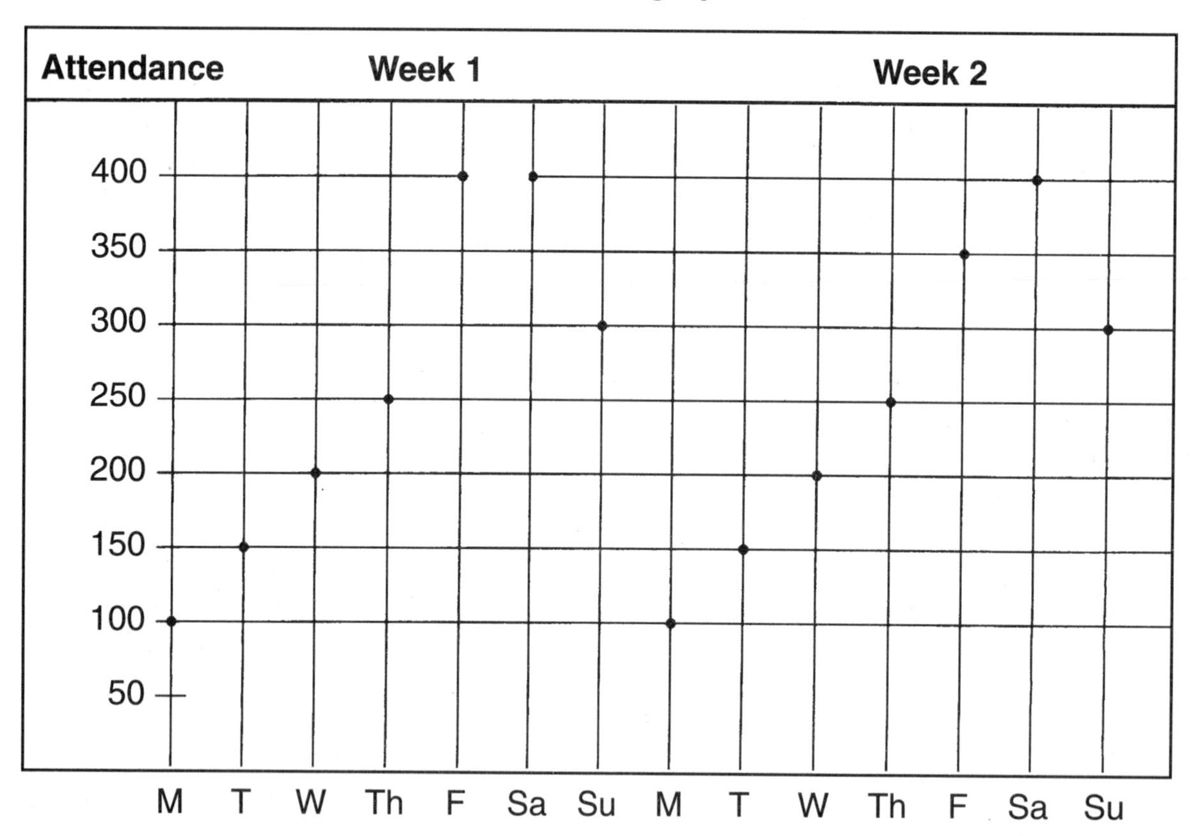

Use the broken-line graph to solve.

1. What was the total number of tickets sold on both Mondays and Tuesdays?

   ____

2. On which night did attendance increase most from the night before?

   ____

3. How many more tickets were sold on the first Sunday than on the second Sunday?

   ____

4. The manager does not work on the two nights each week when attendance is lowest. On which two nights does he not work?

   ____

Name ______________________ Date ______________________

GEOMETRY

Unit 7: Estimating Area

Area is written in *square* units of measurement.
Janice is planning to wallpaper the rooms in her house. To buy the correct amount of wallpaper, she estimates the area of the walls in each room.

1. One wall in Janice's bedroom is 2.89 meters wide. Round this number to its greatest place.

   ____

2. The same wall is 3.75 meters high. Round this number to its greatest place.

   ____

3. Estimate the area in square meters of one wall in Janice's bedroom.

   ____

4. If all 4 bedroom walls have the same area, estimate the total area of the walls.

   ____

5. Each wall in Janice's study is 2.44 meters wide and 3.05 meters high. Estimate the area of the 4 walls in the study.

   ____

6. Two walls in Janice's living room are 3.84 meters wide, and the other two are 4.62 meters wide. If her living room is as high as her bedroom, estimate the area of her living-room walls.

   ____

7. At the store, one roll of wallpaper is 0.75 meters wide and 8 meters long. This is about enough wallpaper to cover one wall in which room?

   ____

8. How many rolls of wallpaper does Janice need to cover the walls in the bedroom?

   ____

9. How many rolls should Janice buy to cover her living-room walls?

   ____

10. Why does Janice buy one extra roll to cover the walls in the bathroom?

   ____

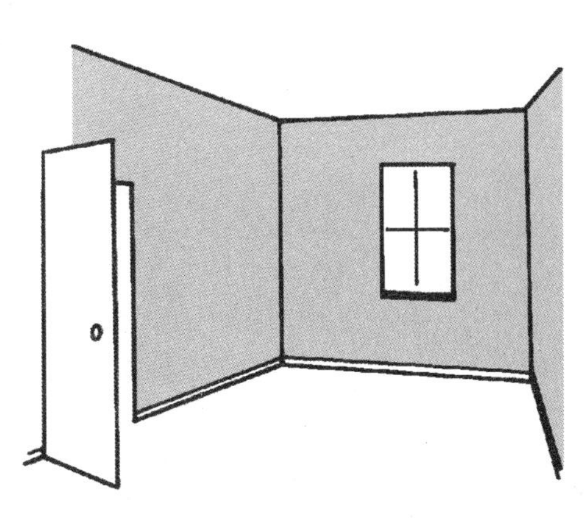

Name ______________________ Date ______________________

GEOMETRY

Fewest Number

Kaws was a territory divided into 5 regions. When Kaws became a state, it was divided into 14 unusually shaped counties.

1. Color the map of the territory so that no two regions of the same color touch each other. What is the fewest number of colors you can use?

   ____

2. Color the larger map of the state so that no two counties of the same color touch each other. What is the fewest number of colors you can use to complete the larger map?

   ____

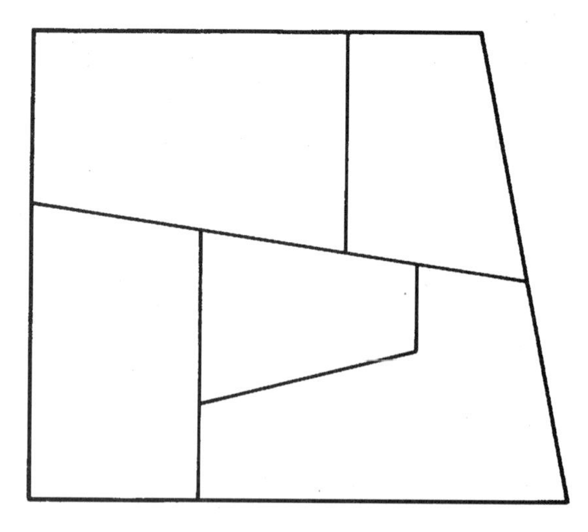

Territory of Kaws

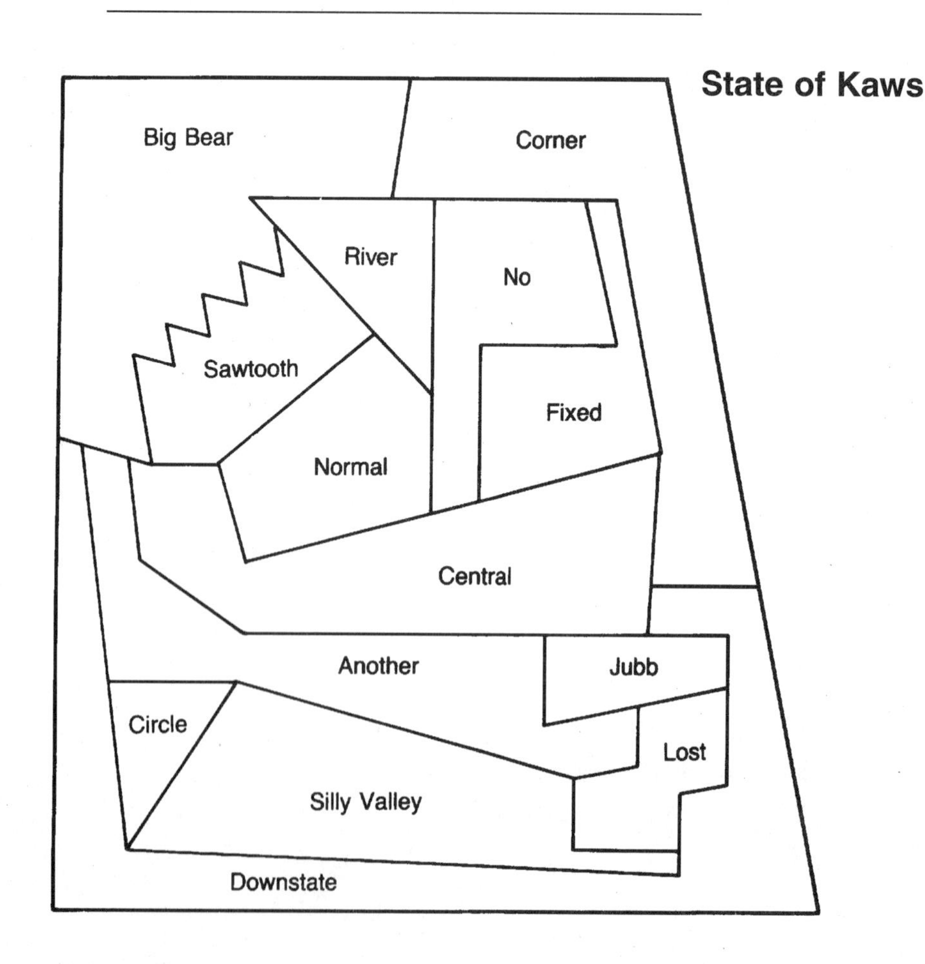

State of Kaws

Name ______________________ Date ______________________

GEOMETRY

Fractions

Complete each equation sentence by writing the letter for the correct figure. Some letters may be used more than once.

1. 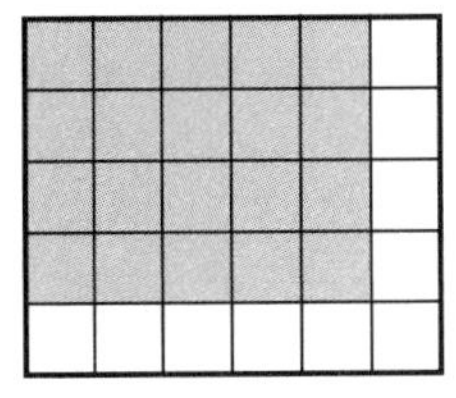 $= 2\frac{6}{7}$ of ________

2. 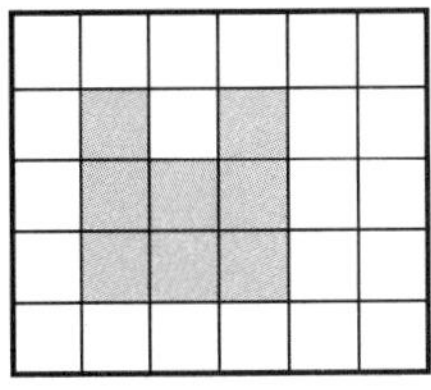 $= 1\frac{3}{5}$ of ________

3. 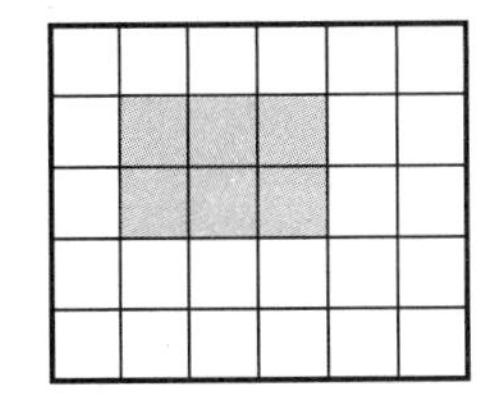 $= 1\frac{1}{2}$ of ________

4. 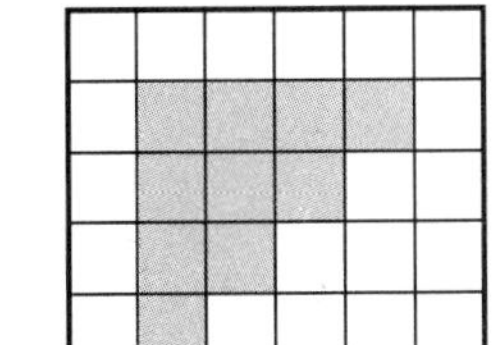 $= 1\frac{1}{9}$ of ________

5. 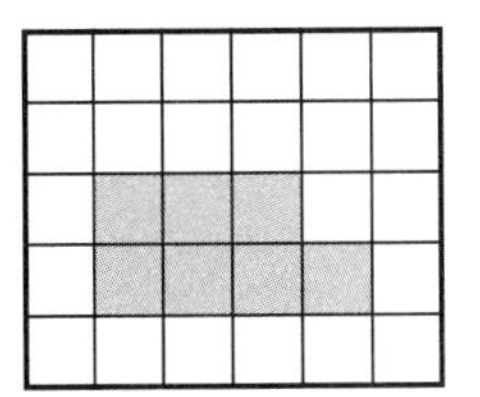 $= 1\frac{3}{4}$ of ________

6. 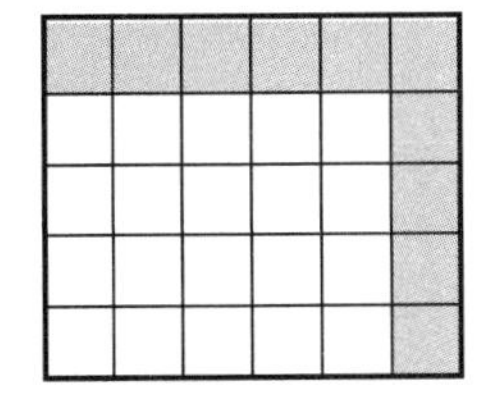 $= 1\frac{3}{7}$ of ________

7. 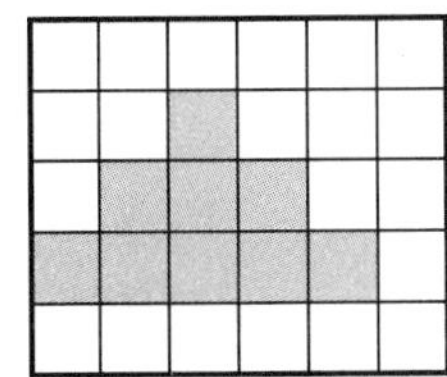 $= 1\frac{1}{8}$ of ________

8. 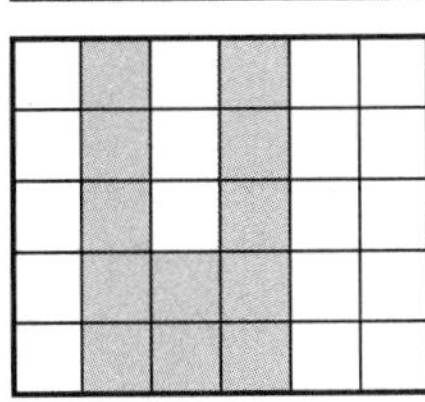 $= 1\frac{1}{11}$ of ________

a. 

b. 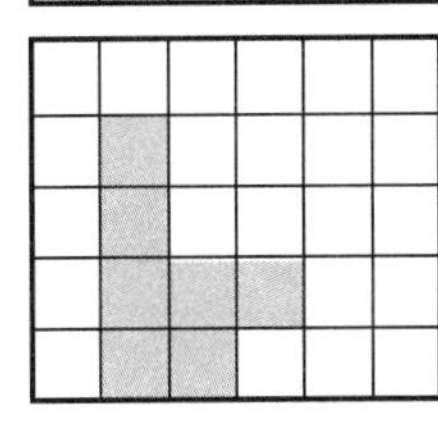

c. 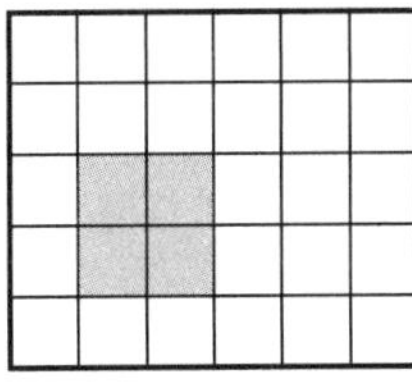

d. 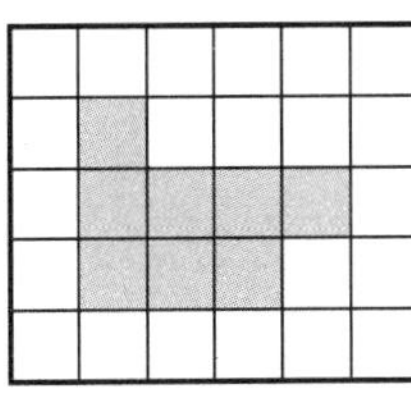

e. 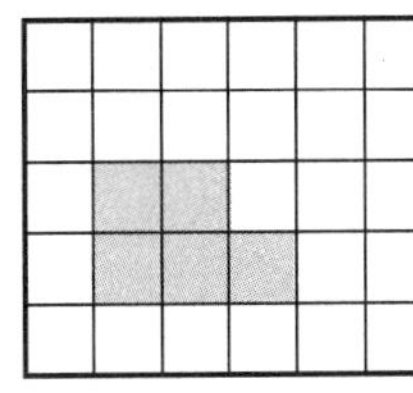

f. 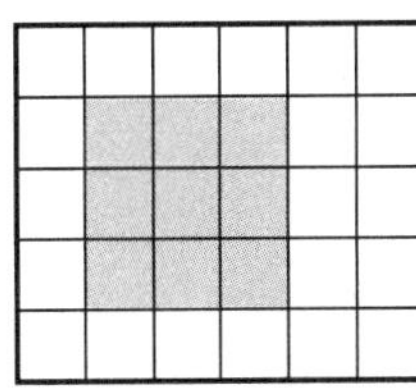

Name ______________________ Date ______________________

GEOMETRY

Basic Ideas of Geometry

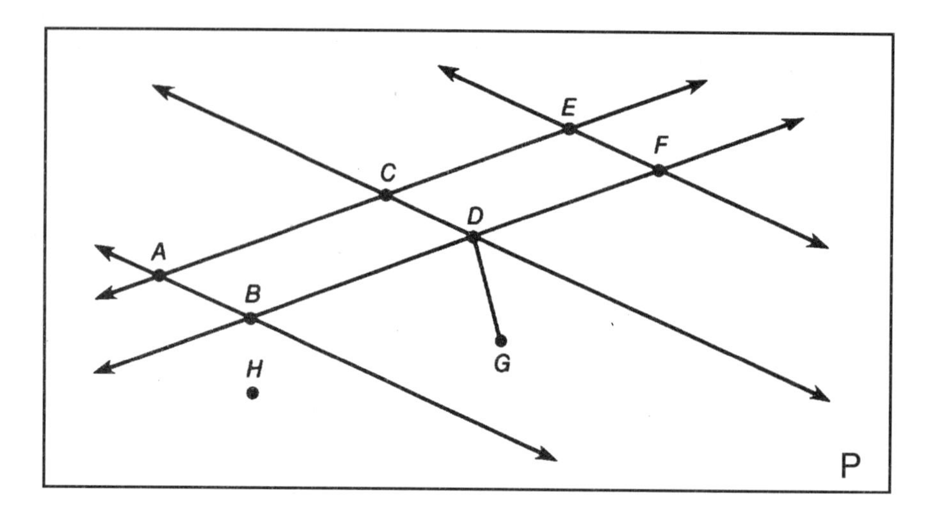

Use the figure to identify each example.

1. point

____

2. line segment

____

3. line

____

4. parallel lines

____

5. intersecting lines

____

6. plane

____

Draw a line to match each item with its geometric description.

7. train tracks
8. a road map
9. a city on a road map
10. the road between two cities on a road map

a. point
b. line seqment
c. parallel lines
d. plane

Name ______________________ Date ______________________

GEOMETRY

Angles

In his geometry class, Darrell learned about the four different kinds of **angles:** acute, obtuse, right, and straight. To help himself remember the difference, he pictured clocks. For instance, Darrell remembered that when school began at 9:00, the hands of the clock formed a right angle.

Write the kind of angle that the hands of each clock make.

1. 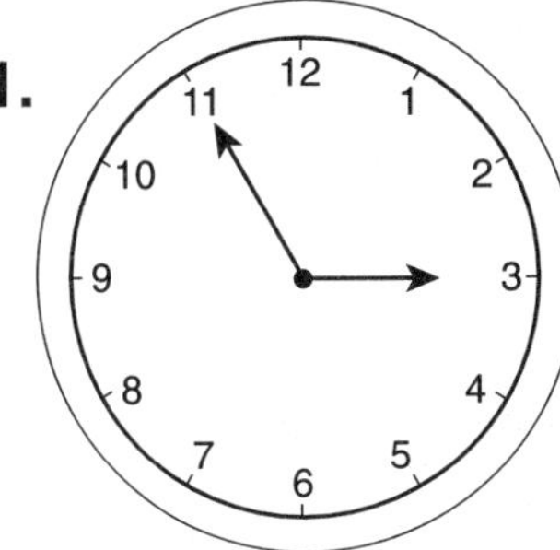

____

2. 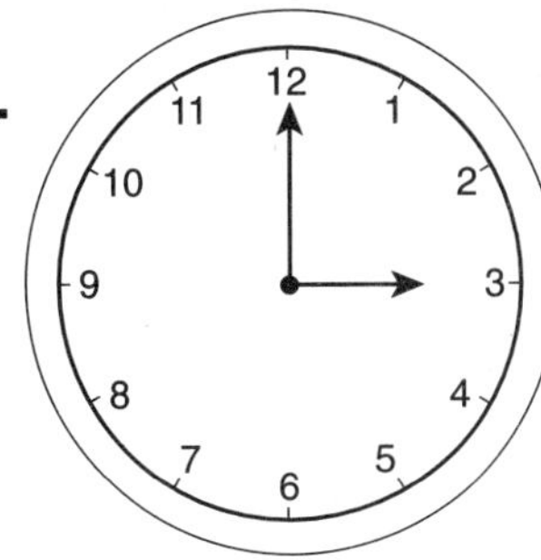

____

3. 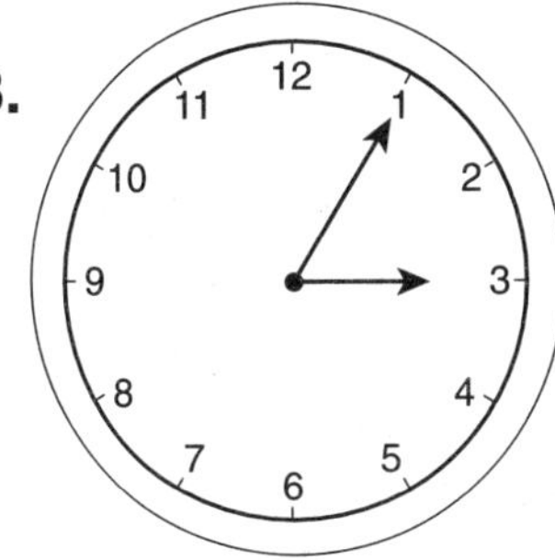

____

4. Darrell is careful to remember that 3:30 is not a right angle. Why is 3:30 not a right angle?

____

5. At about how many minutes after 10:00 do the hands of the clock form a right angle?

____

6. At about how many minutes after 7:00 do the hands of the clock form no angle?

____

7. At about how many minutes before 5:00 do the hands of the clock form a straight angle?

____

What angle do the hands of the clock form when Darrell

8. wakes up at 7:30? ______________________

9. eats lunch at 12:00? ______________________

10. finishes dinner at 6:30? ______________________

Name ______________________ Date ______________________

GEOMETRY

Measuring Angles

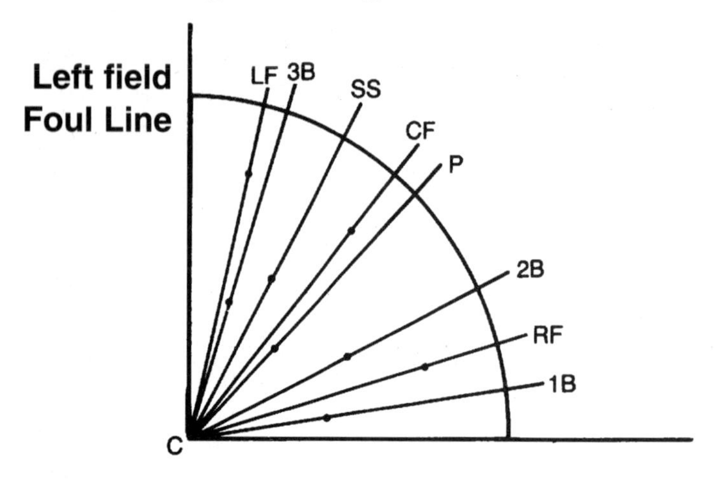

On the baseball diamond diagramed above, the two foul lines form a right angle with home plate as the vertex. The points indicate the position of each of the 9 players in the field.

Measure the angles with your protractor. The right-field foul line is 0°. Write the number of degrees each player is away from the line.

1. the pitcher (P) ______________

2. the first baseman (1B) ______________

3. the second baseman (2B) ________

4. the shortstop (SS) ______________

5. the third baseman (3B) __________

6. the left fielder (LF) ______________

7. the center fielder (CF) __________

8. the right fielder (RF) ______________

9. Which player is closest to the right-field foul line in degrees? ________

10. Which player is closest to the left-field foul line in degrees? ________

11. A left-handed hitter comes up, so the manager shifts his outfield. The left fielder moves to 68°, the center fielder moves to 45°, and the right fielder moves to 10°. Draw this new outfield defense on the diagram below.

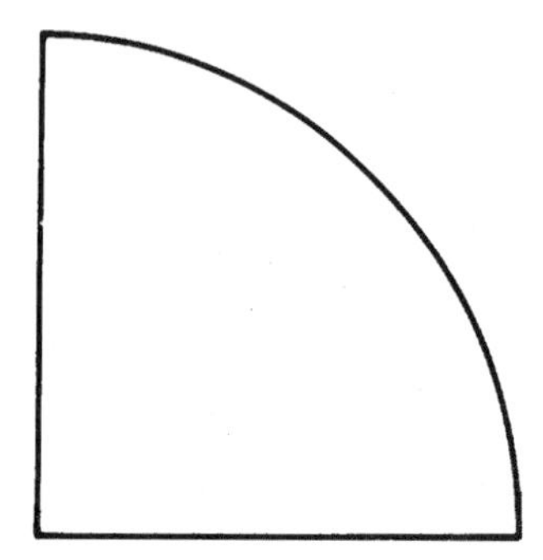

Name ______________________ Date ______________________

GEOMETRY

Triangles

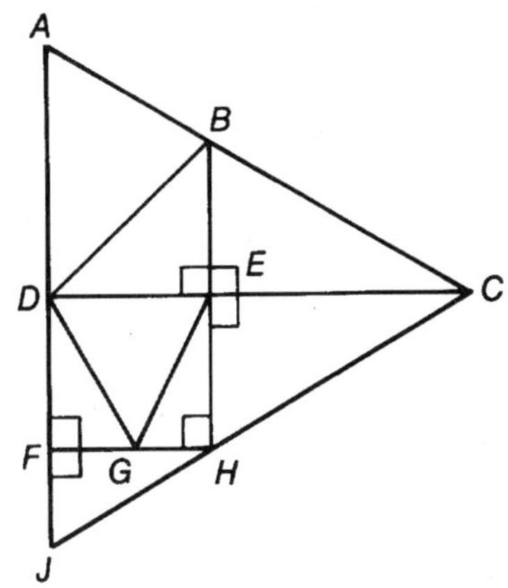

$\overline{BH}$ is perpendicular to $\overline{DC}$, $\overline{HF}$ is perpendicular to $\overline{AJ}$, $\overline{AJ}$ is parallel to $\overline{BH}$.

The following angles have the same measure: $\angle DAB$, $\angle EBC$, $\angle BCH$, $\angle CHE$, $\angle HJF$, $\angle DGF$, $\angle DGE$, $\angle EGH$, $\angle EDG$, and $\angle DEG$.

$\triangle$BED is an isosceles triangle.
The measure of each angle of an equilateral triangle is always 60°.
The measures of the angles opposite the equal sides of an isosceles triangle are also equal.

1. Fill in all 24 angle measurements on the diagram above.

Use the diagram and the information to solve.

2. What are the three equilateral triangles?

____

3. What are the six right triangles?

____

If *BF* were drawn,

4. would $\triangle BFH$ be *right, acute*, or *obtuse*?

____

5. would $\triangle BDF$ be *right, acute*, or *obtuse*?

____

Name ______________________ Date ______________________

GEOMETRY

Mirror Images

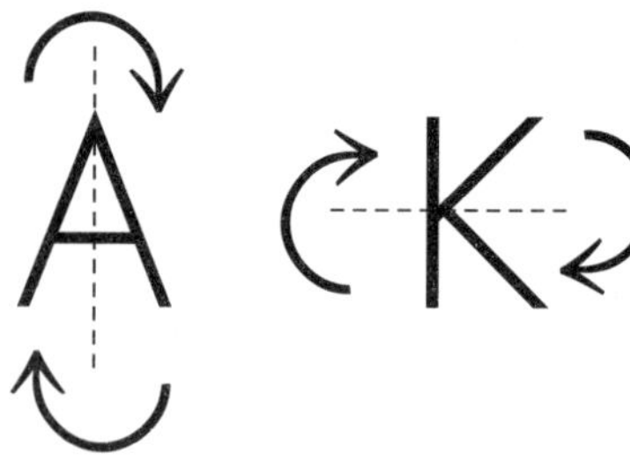

Some letters of the alphabet can be divided by a line so that the parts on each side of the line look like mirror images of each other. Imagine the letter A with a vertical line drawn through it. Each half of the letter is a mirror image of the other half.

The same thing is true of letters that can be divided by a horizontal line. If the letter is divided along that line, each side looks like a mirror image of the other.

Underline the words in the box for which each letter in the word can be divided into mirror images by a vertical line.

HUM	MAT	HIM	RIVER	RUN	CHILD	YAM
VAT	HOLD	MOUTH	HOAX	IVY	CAT	
WAX	DESK	MOW	SEED	MOAT	HIGH	

Look at the words you underlined. List the letters that form mirror images when divided by a vertical line. ______________________

Underline the words in the box that form mirror images when divided by a horizontal line.

BEAD	BECK	READ	COX	KICK	INTO
RACE	CHOKE	DAM	BIKE	HUM	DOCK
CHEER	CALM	CHICK	BAKE	KID	EAR

Look at the words you underlined. List the letters that form mirror images when divided by a horizontal line. ______________________

Underline the words in the box that form mirror images when each letter is divided by a vertical and a horizontal line.

PUSH	SHIN	GRASS	HOPE	SIX	HO
SICK	HIS	RACE	MOON	DAY	ZOO
USE	CLIP	OX	SOON	ON	EAT	HI

Look at the words you underlined. List the letters that form mirror images when divided horizontally or vertically. ______________________

Name ______________________ Date ______________________

GEOMETRY

Network

You can connect the four cities on the map with a series of roads in which *none* of the roads cross.

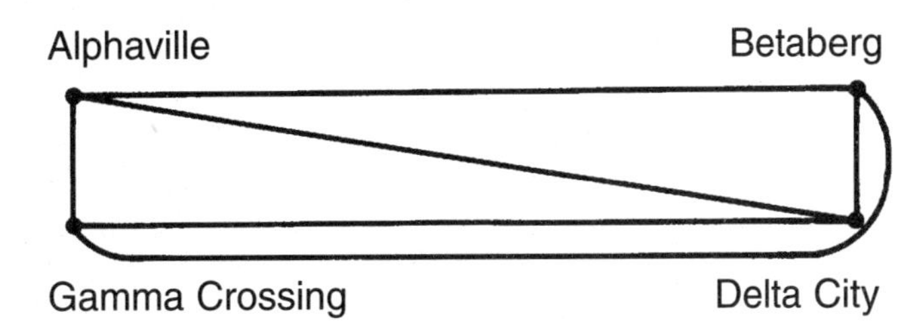

The figure is called a **network**. A *network* is a figure made of *points* (the four cities) and the *arcs* (the roads). An arc may be curved like this: ⌒ ,or it may be a straight line segment like this — :

Look at the figure and answer the questions.

1. Name the points in the network. __________
2. How many arcs connect A and B? __________
3. How many arcs connect B and C? __________
4. How many arcs connect C and D? __________

Look at each network. Write the number of points in each network. Write the number of arcs in each network.

5. 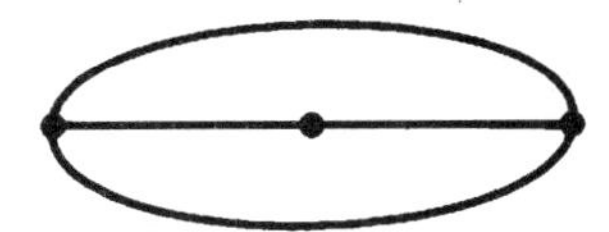

points: __________ arcs: __________

6. 

points: __________ arcs: __________

7. 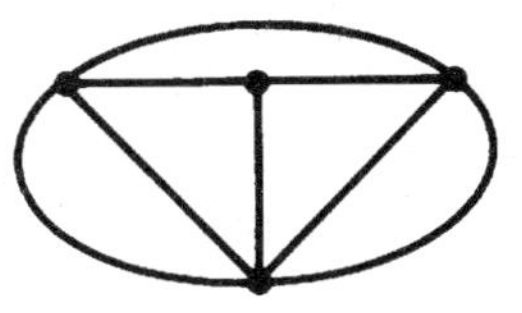

points: __________ arcs: __________

Name ______________________ Date ______________________

GEOMETRY

Perimeter

Look at the figure.

ADGJ is a square.
QOMK is a square.
ATD, *AQB*, and *GUE*
are isosceles right triangles.

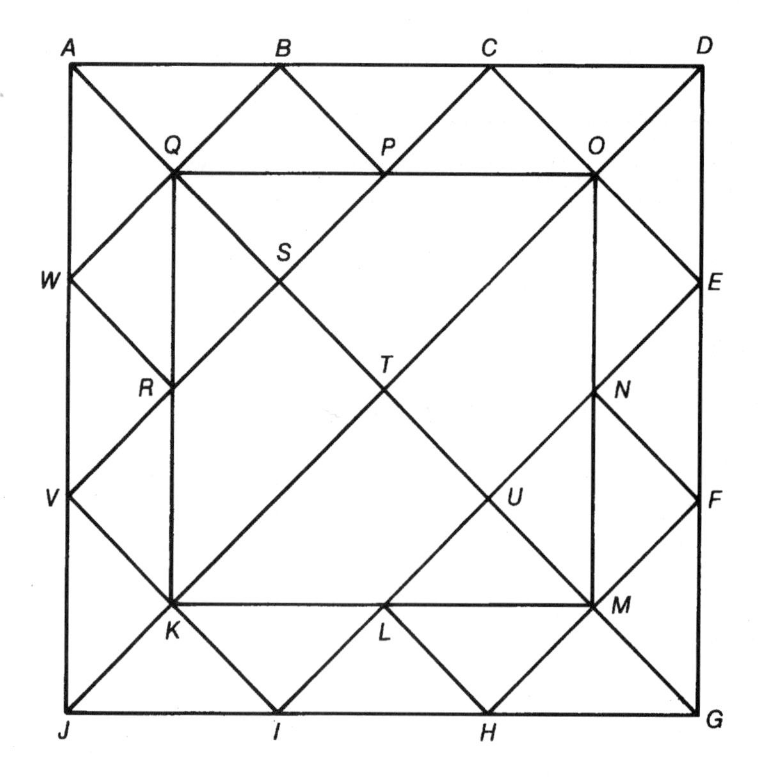

Use the information given about the figure to write the answers.

1. There are seven squares inside *ADGJ*. What are they?

____

____

2. There are eight isosceles right triangles inside *ATD*. What are they?

____

____

3. There are nine rectangles in the figure *ADGJ* that are not squares. What are they?

____

____

4. There are seven parallelograms in *JTG* that are not squares. What are they?

____

____

5. There are seven trapezoids in *ATD*. What are they?

____

____

6. What is the name of the figure *LKRSU*? What is the name of the figure *LKRPON*?

____

____

Name ______________________ Date ______________________

GEOMETRY

Rotating Figures

Look at the letter *D*. Put the point of your pencil in the center of the *D* and rotate the page clockwise a quarter of a circle at a time.

original position 1/4 circle 1/2 circle 3/4 circle full circle

Look at the letters in Exercises 1 through 9. The first letter is the original position of the letter. On the line, write if the letter has been rotated $\frac{1}{4}$, $\frac{1}{2}$, $\frac{3}{4}$, or full circle.

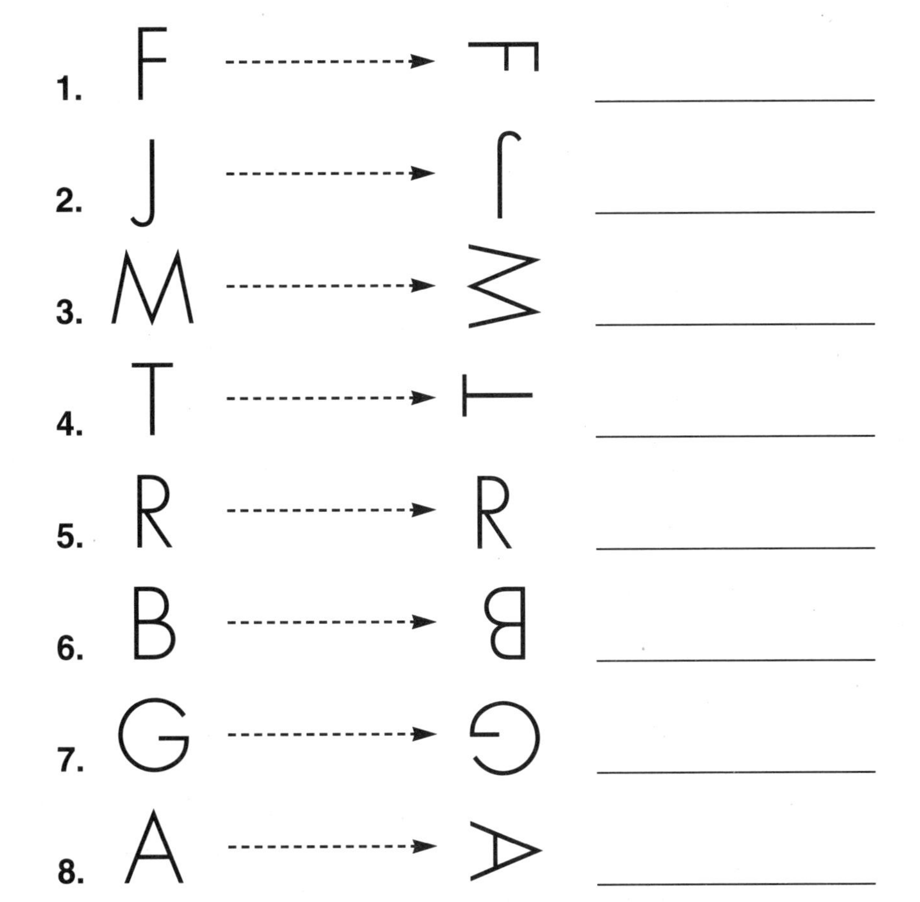

1\. ______________

2\. ______________

3\. ______________

4\. ______________

5\. ______________

6\. ______________

7\. ______________

8\. ______________

9\. ______________

10. Which letter in the alphabet looks the same at $\frac{1}{4}$, $\frac{1}{2}$, $\frac{3}{4}$, and full circle?

____

Name ______________________ Date ______________________

GEOMETRY

Solid Figures

Try to visualize what these three-dimensional forms would look like if they were made from paper and then unfolded.

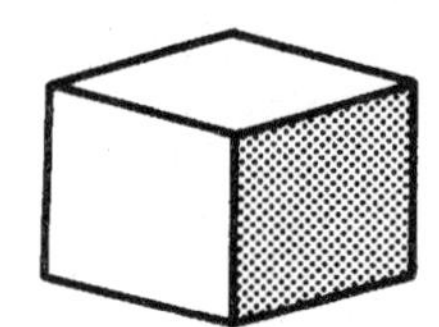 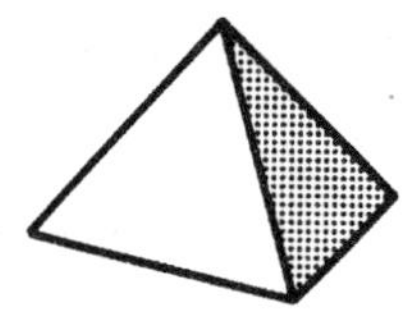 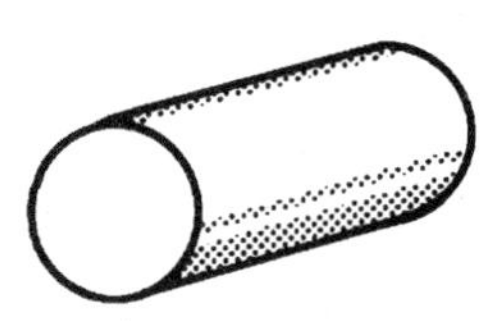

An unfolded cube might look like this. Can these patterns be folded to become cubes? Write *yes* or *no*.

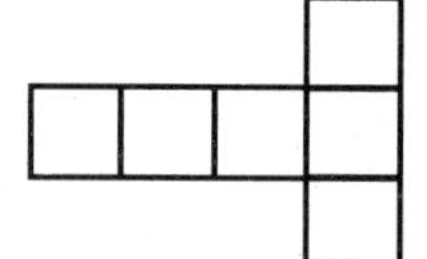

1. __________

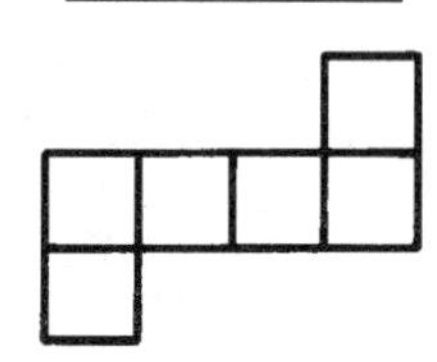

2. __________

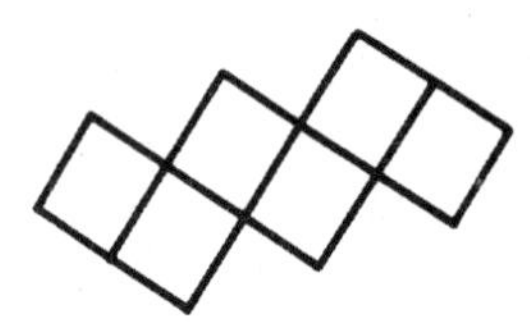

3. __________

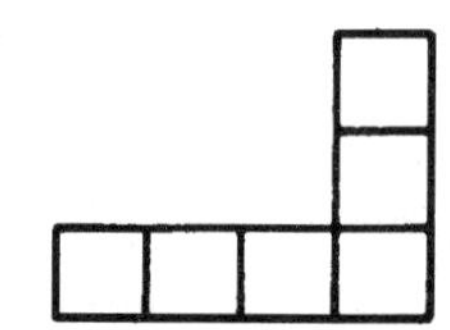

An unfolded rectangular pyramid might look like this. Can these patterns be folded to become rectangular pyramids? Write *yes* or *no*.

4. __________

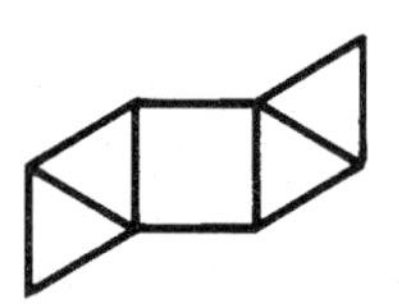

5. __________

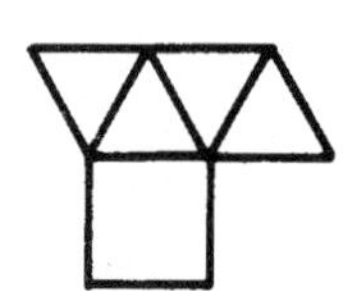

6. __________

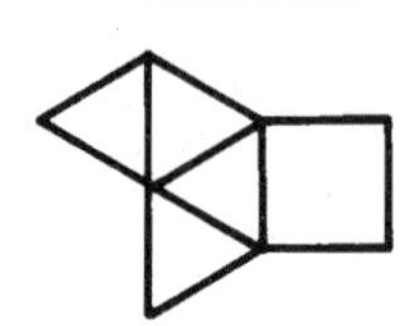

An unfolded cylinder might look like this. Can these patterns be folded to become cylinders? Write *yes* or *no*.

7. __________

8. __________

9. __________

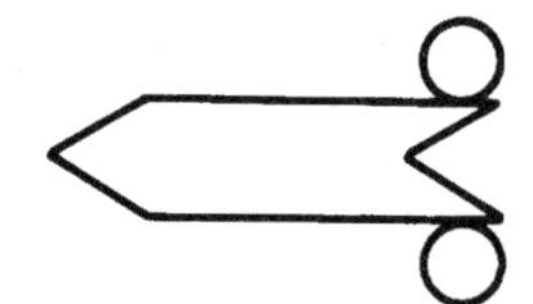

Name ______________________________ Date ______________________________

Unit 8: Sample

You've seen television commercials where they say, "Four of every five people surveyed prefer our product." These claims are based on questions that were asked of many people. This group of people is called a **sample**. It is impossible to ask everyone if they like a product, so samples are used. From the sample, a prediction of likes and dislikes can be made. The larger the sample group, the more accurate the results of the survey will be.

Souper Soup Company thinks that its New Carrot Soup will be very popular. The company took a survey of its employees to see what people thought about the soup.

1. They asked 4 of the 5,000 employees. All 4 loved the soup. The company said that everyone loved the soup. Was this a good sample? Explain.

____

____

____

2. Next, the company surveyed 1,000 of its employees, and 900 said they loved New Carrot Soup. The company decided that 9 of every 10 people would love the soup. Was this more accurate than the other sample? Explain.

____

____

In order to obtain a good sample, a variety of people should be surveyed. The people at Souper Soup may like the product just because they work for the company and feel loyal towards its products.

3. Souper Soup surveyed 1,000 employees of the Simmering Soup Company, a rival soup maker. New Carrot Soup was not a hit. Only 25 people liked the soup; 975 hated it. What was wrong with the sample?

____

____

4. What would be a good sample group for Souper Soup to use to test New Carrot Soup?

____

____

Name ______________________ Date ______________________

Ratio

The fifth-grade class is recording the scores of all the participants in the Ring-Toss game at the school carnival. They drew a table to list the scores of the top four players. Use the table to answer the questions.

The Ring Toss

Players	Total throws	Hits			Total hits	Misses
		Bull's eye	First ring	Second ring		
Sam	20	2	4	10	16	4
Shirley	18	3	6	9	18	0
Jim	24	2	8	8	18	6
Paula	30	5	6	12	21	9

1. What is the ratio of bull's eyes to total throws for each player?

   ____

2. What is the ratio of Jim's hits to misses? Is this ratio greater or less than Paula's ratio of hits to misses?

   ____

3. What is the ratio of Sam's first-ring hits to his second-ring hits? How is this ratio expressed as a percent?

   ____

4. What proportion of Paula's total throws were hits? What proportion were first-ring hits? Write these numbers as percents.

   ____

5. Complete the chart.

Players	Total throws	Hits	Misses	Ratio of hits to misses	Percentage of hits
Claire	25	10	15	2:3	40%
Stefan	20	5			
Jackie	50		30		
Hal	10				80%
Iris	25	15			

Name ______________________ Date ______________________

Reading a Table

Toshi is writing a report about tennis players. She went to the tennis court and asked the players questions. When she had enough information, Toshi made a table.

TENNIS TABLE

Name	G (Girl) B (Boy)	Years playing	Age	Name	G (Girl) B (Boy)	Years playing	Age
Jaqueline	G	5	11	Enrico	B	3	12
Roscoe	B	4	10	Lynn	G	1	10
Shivaji	B	1	11	Al	B	5	11
Sabrina	G	3	14	Juan	B	2	13
Roberto	B	2	13	Angela	G	5	12
Stephanie	G	5	11	Joanne	G	1	9
Lily	G	2	10	David	B	4	11

Use the information from the table to write each answer.

1. Write the ages of all the players from the youngest to the oldest.

____

2. List the number of years each person has played from the least to the most.

____

3. How many girls did Toshi question? How many boys did she question?

____

4. What is the age difference between the youngest player and the oldest player?

____

5. To the nearest year, what is the average age of all the players?

____

6. What number of years appears most frequently under *Years playing*?

____

7. Add the number of years each player has played. Divide by 14. What is the average number of years the people have been playing tennis, to the nearest year?

____

8. What is the age difference between the oldest boy and the youngest girl? What is the age difference between the oldest girl and the youngest boy?

____

____

Math Enrichment:

Assessments

P. 9
1. b, 2. c, 3. a, 4. c

P. 10
1. a, 2. a, 3. c, 4. d

P. 11
1. b, 2. d, 3. d, 4. a

P. 12
1. b, 2. a, 3. c, 4. b

UNIT 1: Number

P. 13
Answers may vary. Accept any reasonable estimate.

Raoul	Model	Estimate	Actual
	XG-1	$400	$351
	TRI-7	$300	$344
	L-69	$60	$61
	L-4	$70	$73
	CC-32	$400	$387
Sum		$1,230	$1,216

John	Model	Estimate	Actual
	X-7	$100	$97
	TRI-7	$300	$344
	L-69	$60	$61
	L-4	$70	$73
	CC-86	$200	$209
Sum		$730	$784

Mabel	Model	Estimate	Actual
	XL-7	$500	$511
	TRI-7	$300	$344
	L-4	$70	$73
	L-69	$60	$61
	CC-44	$100	$148
Sum		$1,030	$1,137

Cherise	Model	Estimate	Actual
	XL-7	$500	$511
	TRI-3	$200	$215
	L-80	$40	$42
	L-69	$60	$61
	CC-44	$100	$148
Sum		$900	$977

Cherise won.

P. 14
1. $0.39, 2. $2,558, 3. $10,662, 4. $2,951, 5. $8,110, 6. 4,251, 7. $12,800, 8. $1,150, 9. $1,290, 10. $39,342

P. 15
Answers will vary. Possible answers:
1. Start – 0.003 – 0.092 – 0.093 – 0.11 – 0.112 – End
2. Start – 0.003 – 0.073 – 0.092 – 0.093 – 0.11 – 0.112 – 0.32 – End
3. Start – 0.003 – 0.092 – 0.11 – 0.112 – End;
Start – 0.003 – 0.05 – 0.073 – 0.112 – End;
Start – 0.04 – 0.1 – 0.13 – 0.32 – End;
Start – 0.04 – 0.05 – 0.112 – 0.32 – End

P. 16
1. 7; 5, 2. 35; 3; 5, 3. 88; 88; 8,835
4. 102, 17; 77, 6; 77, 77; 78; 7,802
5. 7,178; 26 x 3 = 78; 74 - 3 = 71; 97 - 26 = 71
6. 8,722; 11 x 2 = 22; 89 - 2 = 87; 98 - 11 = 87
7. 7,968; 4 x 17 = 68; 96 - 17 = 79; 83 - 4 = 79
8. 7,912; 8 x 14 = 112; 86 - 8 = 78; 9 = 78; 78 + 1 = 79

P. 17

1. 5 and 10	2. VII and XII
3. XIII	4. XX
5. XVI	6. XI
7. XXXVII	8. XXV
9. XII	10. XXVIII
11. XXVIII	12. XXXVI
13. LXVII	14. LXXVIII
15. CXLI	16. XCVIII

P. 18
1. 15, 2. 67.5, 3. 67.5, 4. 135, 5. 2, 6. 9
7. 8.

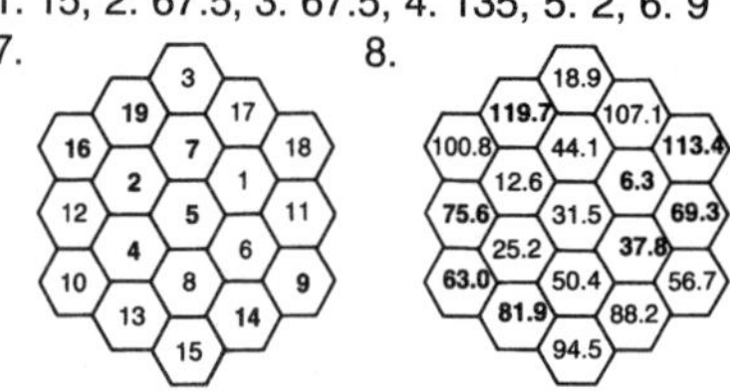

9. 38, 10. 239.4

p. 19
1. 4,020
+ 938
4,958

2.

×	300	20	5	
20	6,000	400	100	6,500
5	1,500	100	25	+ 1,625
				8,125

3.

×	200	70	5	
10	2,000	700	50	2,750
4	800	280	20	+ 1,100
				3,850

4.

×	100	70	3	
90	9,000	6,300	270	15,570
1	100	70	3	+ 173
				15,743

5.

×	3,000	900	20	9	
10	30,000	9,000	200	90	39,290
2	6,000	1,800	40	18	+ 7,858
					47,148

6.

×	4,000	100	60	2	
900	3,600,000	90,000	54,000	1,800	3,745,800
90	360,000	9,000	5,400	180	374,580
9	36,000	900	540	18	+ 37,458
					4,157,838

7.

×	2,000	200	7	
300	600,000	60,000	2,100	662,100
30	60,000	6,000	210	66,210
0	0	0	0	+ 0
				728,310

P. 20
Order of maze:
North Gate to 2.64 ÷ 4 = 0.66 to 2.994 ÷ 3 = 0.998 to 77.92 ÷ 8 = 9.74 to 86.037 ÷ 7 = 12.291 to 98.82 ÷ 61 = 1.62 to 9.044 ÷ 17 = 0.532 to 89.480 ÷ 20 = 4.474 to 69.027 ÷ 7 = 9.861 to South Gate

P. 21
1. 3, 2. 17, 3. 38, 4. 15, 5. 16, 6. 33, 7. 13
8. 27

P. 22
1. Answers may vary. Possible answers:

6	=	3	+	3
8	=	3	+	5
14	=	3	+	11
24	=	17	+	7

2. Answers will vary. Possible answers:

18	=	7	+	11
36	=	5	+	31
44	=	7	+	37
22	=	3	+	19

13	+	5	=	18
7	+	29	=	36
41	+	3	=	44
5	+	17	=	22

3. Answers will vary. Possible answers:

27	=	7	+	17	+	3
53	=	11	+	31	+	11
75	=	3	+	11	+	61

7	+	7	+	13	=	27
5	+	7	+	41	=	53
5	+	11	+	59	=	75

P. 23
Table B: 642, Table C: 256.8; 513.6
Table D: 481.5, Table E: 535
C shows plans that will balance the plane's cargo.
Comp. 1: Answers will vary; must <500.
Comp. 2: Answers will vary; must be <500.
Comp. 3: Answers will vary; must <700.
Comp. 4: Answers will vary: must be <700.
Comp. 5: Answers will vary; must be <700.

P. 24
1. 10 parts, 2. 0.2, 3. 6.2
4. 0.1 and 0.2; 0.3 and 0.4
5. Dots should be accurately placed on number line.
6. twice as great

P. 25
GRAYMOTT'S DOG FOOD
Meat by-product: 19%
Vitamin A: 8%
Water: 13.5 g
DELPHA DOG FOOD
Fat: 64.5 g
Calcium: 10%
Preservatives: 10.5 g
1. Vitamin A: 29.25 g
Calcium: 22.5 g
O'CONNOR'S BIRD SEED

Crushed peanuts	11%	5.5 g	8.8 g
Sunflower seeds	8%	4 g	6.4 g
Sesame seeds	39%	19.5 g	31.2 g
Caraway seeds	42%	21 g	33.6 g

Math Enrichment: GRADE 5

UNIT 2: Problem Solving

P. 26
1. \$8,500, 2. \$11,200, 3. \$5,800
4. \$13,900, 5. \$9,800, 6. \$9,900
7. \$6,600, 8. \$6,500, 9. \$3,800
10. \$3,700, 11. \$7,500, 12. 0

P. 27
1. Students should mark line graph to show 50 million in 1970, 2. Students should mark bar graph to show 150 million in 1970, 3. 20 million, 4. the urban population, 5. 1890 and 1910, 6. 1910
7. no, 8. that they have grown also

P. 28
1. 48.7 - 2.9 + 0.47 = *n*; 46.27 *s*
48.7 + 45.8 + 46.27 = *y*; 140.77 *s* or 2 min 20.77 *s*, 2. 6 min 32 *s* + 2 min 23.3 *s* = *n;* 8 min 55.3 *s*, 3. 1.35 + 0.42 + 0.74 = *n;* 2.51 million copies, 4. 47,320 - 2,420 + 1,755 = *n;* 46,655 fans

P. 29
1. yes, 2. one more roll, 3. no, 4. Ward Hardware, 5. 4, 6. no

P. 30
1. Multiply to find how many miles they walk in 4 days., 2. Subtract to find how many miles farther it is., 3. 20 miles, 4. the Tates, 5. yes, 6. 6 miles

P. 31
1. 57; *n* = 102 - 45, 2. \$8.00; *n* = \$26 - \$18
3. \$208.00; *n* = 4 x \$52, 4. 75; *n* = 225 ÷ 3
5. 9; *n* = 81 ÷ 9, 6. 56; *n* = 11.2 x 5
7. 0.6 hours; *n* = 3 - 2.4
8. 5; *n* = 1.2 + 2.3 + 1.5

P. 32
Answers will vary. These are reasonable answers.

Dish	Estimated cooking time	Starting time	Estimated serving time
Soup	10 minutes	4:50 p.m.	5:00 p.m.
Beef	2 hr 45 minutes	2:25 p.m.	5:10 p.m.
Potatoes	55 minutes	4:15 p.m.	5:10 p.m.
Beans	15 minutes	4:55 p.m.	5:10 p.m.
Dessert	65 minutes	4:40 p.m.	5:45 p.m.

P. 33
1. Draw line through *Her head is 5.18m.*; 11.87 m, 2. Draw line through *The mouth is 0.80 m wide.*; 3 m wide, 3. Draw line through *and has columns 21.8 m high on it*; 91.52 m, 4. Draw line through *The copper covering the statue is 2.3 mm thick.*; 11,250 kg, 5. Draw line through *101,250 kg of steel were used.*; 156,521.73 kg, 6. Draw line through *He died in 1904.*; 1834, 7. Draw line through *the second 2.85 meters*; 11 meters

P. 34
1. 5 leaves, 2. 5 people, 3. 12 spools,
4. \$12.50, 5. \$8.00

P. 35
1. 36; no, 2. 8 full carriers; 2 tanks left
3. 8 groups; 4 pictures; 15 pictures
4. 5 per person; 3 per day; 2 people; 4 experiments, 5. 9 trips
6. 8 times; 3 packs left

P. 36
1. Central, 2. Pacific, 3. 11:00 a.m.
4. Eastern and Central, 5. 3 times
6. Mountain

P. 37
1. 126 reptiles and amphibians
2. 104; 14, 3. 20% birds, 4. small predators; 7 more, 5. 40% reptiles; 56 reptiles, 6. 42; more, 7. endangered species; 7 more birds

P. 38

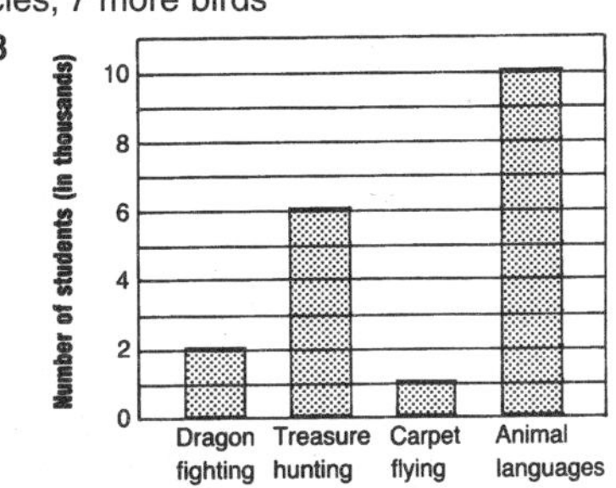

1. 900 students, 2. carpet flying, 3. 300 students, 4. 9,100 more people, 5. the college, 6. one of the longest

UNIT 3: Logic

P. 39
1. 8 nickels, 4 dimes, 2 quarters
2. 39 dimes, 60 pennies, 3. 36¢
4. 10 coins: one half-dollar, one quarter, 4 dimes, 4 pennies, 5. 2 quarters, 10 nickels, 20 dimes, 6. 4 quarters, 4 dimes, 12 nickels, 7. 32 nickels, 10 pennies, 8 dimes

P. 40
Answers will vary. Possible answers:
1. No numbers greater than 20 are less than 10., 2. No numbers less than 10 are greater than 20., 3. Some numbers less than 10 are even numbers.,4. Not all numbers less than 10 are even numbers.
5. Outside circle is Even Numbers, inside circle is Multiples of 10.

P. 41
Answers will vary.

P. 42
1. one, 2. horse or pirate, 3. bunny or toga
4. no; he must have the same costume as Eric, who isn't a pirate

	Horse	Horse	Bunny	Toga	Pirate	Pirate
Marcus	X a.	X a.	O c.	X c.	X a.	X a.
Cindy	X a.	X a.	X c.	O c.	X a.	X a.
Eileen	X b.	X b.	X a.	X a.	O b.	X b.
Steve	O b.	X b.	X a.	X a.	X b.	X b.
Eric	X b.	O b.	X a.	X a.	X b.	X b.
Meredith	X b.	X b.	X d.	X b.	X b.	O b.

Marcus: bunny; Cindy: toga; Eileen: pirate
Steve: horse; Eric: horse; Meredith: pirate

P. 43
1. Flyers, 2. Panthers, 3. Busters,
4. Kangaroos, 5. Mustangs, 6. Hornets,
7. Cougars, 8. Sharks

P. 44
1. Jack: 30; Martha: 20, 2. Sue: 8; Nancy: 5; Jane: 3, 3. Alexandra: 6; Juanita: 16; Carleton: 21; Emilio: 26, 4. Brad: 29 years; Manuel: 38 years 8 months, José: 25 years 10 months; Douglas: 38 years 9 months
5. Yukio: 5 years 7 months; Father: 33 years 6 months, Mother: 27 years 11 months

P. 45
1. 2 places, 2. by knowing that he does not teach languages
3.

	Math	History	French	Spanish
Mr. Richards	X	X	O	O
Mrs. Bellamy	X	O	X	O
Mr. Jimenez	O	O	X	X
Ms. Eshghi	O	X	O	X

4. Mr. Richards: New York; Mrs. Bellamy: Vermont, Mr. Jimenez: California; Ms. Eshghi: Idaho

P. 46
1. 21/25, 2. 2/3, 3. 25/32, 4. 25/48
5. no; 19/27, 6. Eileen's; 25/64, 7. Buzz; Donna; Ed; Manuel; Celia; Eleanor; Eileen

P. 47
1.

Name	Time
Annie	10:00 p.m.
Jamal	9:40 p.m.
Li	10:20 p.m.
David	11:05 p.m.
Enid	10:15 p.m.
Fred	10:35 p.m.
Jorge	11:35 p.m.

2.

Name	Time
Annie	6:25 a.m.
Jamal	6:05 a.m.
Li	7:05 a.m.
David	7:00 a.m.
Enid	7:05 a.m.
Fred	7:30 a.m.
Jorge	8:30 a.m.

3. no one, 4. David, 5. David, 6. Fred and Jorge, 7. Enid, 8. Fred and Jorge

P. 48
1. Box 1: oranges; apples and oranges
Box 2: apples; apples and oranges
Box 3: apples; oranges, 2. The box contains only apples, 3. The box's label is wrong; so, it can't have apples and oranges. The list from Exercise 1 proves this., 4. Apples and oranges. It can't contain only oranges (because the box is mislabeled) or only apples (which are in box 3)., 5. Oranges; that's all that's left.
6. Box 1: apples and oranges; Box 2: apples.

UNIT 4: Patterns

P. 49
1. 8, 14;, 2. 4 1/2, 1 1/2;,
3. 3, 4.5, 4. 19, 29 1/8
5. 54, 162, 486, 6. 702.4, 2,809.6
7. 15 3/16, 22 25/32, 8. 3.75, 3.375, 3.1875

P. 50

Number	Sum of digits	Divisible by 3?	Quotient
1.	6	yes	662
2.	4	no	
3.	9	yes	15,062,670
4.	4	no	
5.	8	no	
6.	2	no	
7.	3	yes	134,494
8.	5	no	
9.	5	no	
10.	3	yes	1,324,495,207

Math Enrichment: GRADE 5

P. 51

1. 2,403, 2,406, 2,409, 2,412, 2,415, 2,418, 2,421, 2,424, 2,427
801, 802, 803, 804, 805, 806, 807, 808, 809
2. yes, 3. Each is 3 greater than the number to the left of it., 4. Each is 1 greater than the number to the left of it.
5.

÷	31,224	31,212	31,200	31,188
2	15,612	15,606	15,600	15,594
3	10,408	10,404	10,400	10,396
4	7,806	7,803	7,800	7,797
6	5,204	5,202	5,200	5,198

÷	32,412	32,400	32,388	32,376
2	16,206	16,200	16,194	16,188
3	10,804	10,800	10,796	10,792
4	8,103	8,100	8,097	8,094
6	5,402	5,400	5,398	5,396

P. 52

X	0	1	2	3	4
0	0	0	0	0	0
1	0	1	2	3	4
2	0	2	4	1	3
3	0	3	1	4	2
4	0	4	3	2	1

X	0	1	2	3	4	5	6	7	8	9
0	0	0	0	0	0	0	0	0	0	0
1	0	1	2	3	4	5	6	7	8	9
2	0	2	4	6	8	0	2	4	6	8
3	0	3	6	9	2	5	8	1	4	7
4	0	4	8	2	6	0	4	8	2	6
5	0	5	0	5	0	5	0	5	0	5
6	0	6	2	8	4	0	6	2	8	4
7	0	7	4	1	8	5	2	9	6	3
8	0	8	6	4	2	0	8	6	4	2
9	0	9	8	7	6	5	4	3	2	1

P. 53

1. $\frac{1}{2}$
2. $3\frac{1}{2}$, 4, $4\frac{1}{2}$
3. $\frac{1}{4}$, $\frac{3}{4}$, $\frac{5}{4}$
4. $\frac{14}{8}$, $\frac{36}{16}$, $\frac{44}{16}$, $\frac{104}{32}$, $\frac{120}{32}$
5. $\frac{8}{6}$, $\frac{24}{12}$, $\frac{56}{24}$, $\frac{64}{24}$
6. You add $\frac{1}{3}$ to the previous term.
7. $\frac{12}{9}$, $\frac{15}{9}$, $\frac{18}{9}$, $\frac{7}{3}$, $\frac{8}{3}$
8. You add $\frac{1}{3}$ to the previous term.

P. 54

1. 212,212, 2. 805,805, 3. 479,479
4. The second factor repeats itself.
5. 2,997, 6. 8,991, 7. 4,995
8. The product of the second factor 9 appears in the thousands and the ones places. The number 9 appears in the hundreds and the tens places.
9. 111,111, 10. 222,222, 11. 444,444
12. Multiplying some multiple of 7 by 15,873 results in a 6-digit product. Each digit is the number by which 7 is multiplied to get the multiple of 7, 13. 333,333, 14. 567,567, 15. 999,999, 16. 7,992

P. 55

1. 284, 2. 28, 3. 18 and 20

P. 56

Note: Students will need tracing paper and scissors.

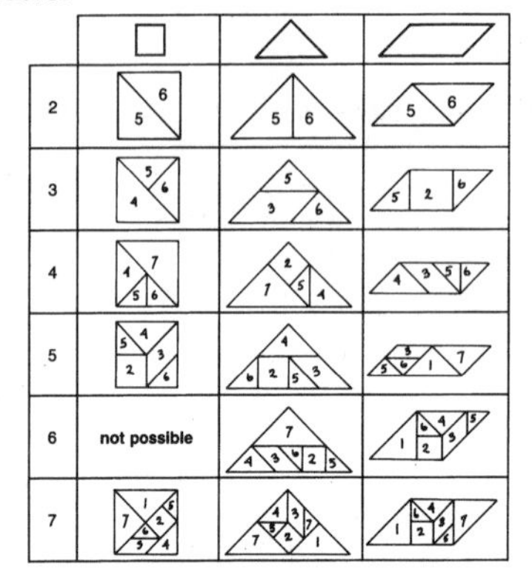

P. 57

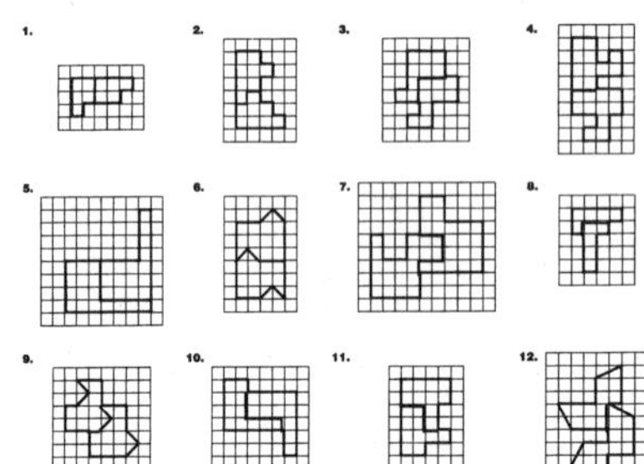

P. 58

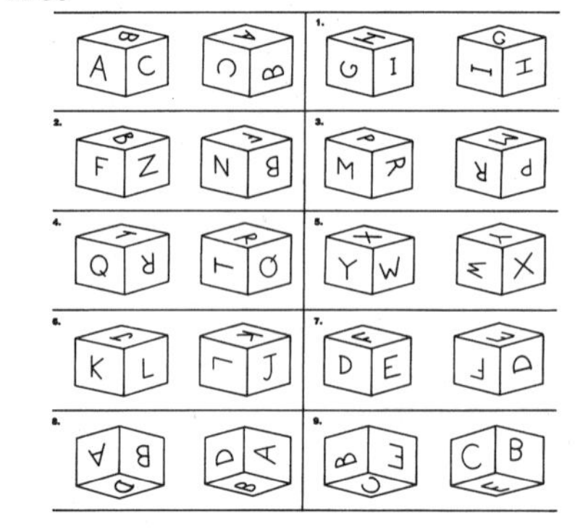

UNIT 5: Algebra

P. 59

1. subtract, 2. add, 3. subtract, 4. subtract, 5. subtract, 6. subtract
Answers may vary. Possible answers:
8. 12 - 3 + 5 = 14, 9. 17 - 16 + 7 = 8
10. 9 - 3 + 1 = 7, 11. 18 - 17 + 3 = 4
12. 9 + 10 - 16 = 3, 13. 13 + 13 - 14 = 12
14. 8 + 10 - 12 = 6

P. 60

1. 1.006	2. 6.89	3. 24.06
+ 0.742	+ 2.46	+ 16.14
1.748	9.35	40.20

4. 3.462	5. 9.358	6. 52.64
+ 0.048	+ 0.947	+ 10.57
3.510	10.305	63.21

Answers may vary in columns with 2 missing digits.

7. 12.9	8. 7.18	9. 1.54
+ 43.8	+ 2.36	+ 6.29
56.7	9.54	7.83

10. 28.1	11. 7.54	12. 21.4
+ 39.4	+ 1.82	+ 65.9
67.5	9.36	87.3

13. Answers will vary. Addition is Commutative.

P. 61

1. 5; Associative, 2. 4; Distributive
3. 1; Identity, 4. 4; Commutative
5. 4; Distributive, 6. 1; Identity
7. 8; Commutative, 8. 2; Distributive
9. 9; Commutative
A: Because it did not add up.

P. 62

1. add, 2. divide, 3. subtract, 4. multiply, 5. multiply, 6. divide, 7. 476, 8. 95, 9. 2,193, 10. 58,608, 11. 521,575, 12. 9

P. 63

Cheese Bake:
$2\frac{2}{3}$ c flour
4 tsp baking powder
$1\frac{1}{3}$ tsp salt
$5\frac{1}{3}$ tsp butter
$1\frac{1}{3}$ c grated cheddar cheese
$\frac{8}{9}$ c cold water

Potato Delight:
7 c bread crumbs
14 eggs
49 oz evaporated milk
$3\frac{1}{2}$ tsp onion powder
$10\frac{1}{2}$ tsp salt
$5\frac{1}{4}$ tbsp Worcestershire sauce
14 potatoes
28 oz frozen vegetables
$17\frac{1}{2}$ raw carrots
$24\frac{1}{2}$ oz gravy

P. 64

Les: 60 inches
Denise: 90
Freida: 104
Luis: 96
Josh: 62 inches
1. 144, 2. 1 inch, 3. 96%; 100%, 4. 6%

UNIT 6: Measurement

P. 65

1. 4.88 trillion, 2. 10.54 light-years, 3. 16.27 light-years, 4. 8.34 light-years, 5. 14.46 light-years, 6. 8.65 light-years, 7. Divide 5.88 by 365 days (1 year); 16.1 billion miles

P. 66

1. 60 lines; 50 lines, 2. 60 lines, 3. no, 4. 52 lines; 65 lines, 5. 900 characters, 6. 280 characters

P. 67

1. 15 inch squares, 2. 27 inch squares, 3. 24 inch squares, 4. 18 inch squares, 5. 60 inch squares, 6. 180 inch squares, 7. 32 inch squares

P. 68

1. 9: 40 a.m., 2. 12:45 p.m., 3. 1:00 p.m.; 3:00 p.m., 4. It will arrive at 11:30 a.m. the following day., 5. 8:55 a.m., 6. Yes; there is a 1-hour wait between the plane's landing in Sarasota and the bus's departure.